Bread from Stones

A
New and Rational System
of
Land Fertilization
and
Physical Regeneration

by **JULIUS HENSEL**

TRANSLATED FROM THE GERMAN
PRICE TWENTY-FIVE CENTS
PHILADELPHIA, PA.
A. J. TAFEL, 1011 ARCH STREET
1894

Edition 2010 by John Schacher

© copyright John Schacher 2010. All rights reserved
© der deutschen Ausgabe: John Schacher, Augsburg 2010
Alle Rechte vorbehalten, insbesondere das der Übersetzung, des öffentlichen Vortrags sowie der Übertragung durch Rundfunk und Fernsehen, auch einzelner Teile.
Kein Teil des Werkes darf in irgendeiner Form (durch Fotografie, Mikrofilm oder andere Verfahren) ohne schriftliche Genehmigung des Verlages reproduziert oder unter Verwendung elektronischer Systeme verarbeitet, vervielfältigt oder verbreitet werden.
1. Auflage 2010

www. julius-hensel.com

ISBN 978-1-4467-5966-0

CONTENT:

PUBLISHER'S PREFACE	6
AUTHOR`S PREFACE	7
THE CAUSE OF THE DECADENCE OF AGRICULTURE	8
HEALTHY AND UNHEALTHY PRODUCE	17
WHAT SHALL WE DO WITH STABLE MANURE?	29
WILL FERTILIZING WITE STONE-MEAL PAY?	35
A CHAPTER FOR CHEMISTS	43
STONE-MEAL AS A TOBACCO FERTILIZER	50
A PAPER CONTRIBUTED TO THE "DEUTSCHES ADELSBLATT"	53
STONE-MEAL MANURE	61
CONTRIBUTIONS FROM OTHER SOURCES	68
STONE-MEAL	68
STONE FERTILIZING	71
LETTER TO MR. SCHMITT	73
TO THE POMOLOGICAL SOCIETY "HEIMGARTEN IN BUELACH"	75
THE STONE-MEAL OF DR. HENSEL BEFORE THE COMMITTEE ON FERTILIZERS OF THE GERMAN AGRICULTURAL SOCIETY	78
ABOUT STONE-MEAL MANURE	82
WHAT HELP CAN BE GIVEN TO THE HARD-PRESSED FARMERS	84
FROM THE "RHEINISCHER COURIER"	86
FROM THE "NEUES MANNHEIMER VOLKSBLATT"	88
"IRON SLAG" FROM "KOELNISCHE VOLKSZEITUNG"	88
"NEUES MANNNHEIMER VOLKSBLATT"	90
"WIESBADENER GENERAL ANZEIGER"	91
DETAILLED STONE-MEAL-EXPERIMENT	94

PUBLISHER'S PREFACE:

healthy, wholesome and life-sustaining; the plants being healthy will escape disease and parasites, and many of the ills of man due to unwholesome food from plants will disappear. Is it not sound reason to believe that food-yielding plants grown on pure, uncontaminated soil will be wholesomer than those grown on soil saturated with sewage and rotting manure from stables?

This is but a brief outline of the theories propounded by Hensel and put in practice in Germany for the last five or six years with amazing success. Put in practice in this country it will not only free the farmer I from a heavy yearly expense for artificial fertilizers but will gradually bring back his exhausted fields to their virgin state and give the public food on which health may be maintained.

In Germany this has become a "cause" sustained by enthusiastic supporters not only among farmers, horticulturalists, florists and gardeners, but also among clergyman, physicians and public-spirited men. They see in it one of the means by which the human race is to be at least physically regenerated, and a sound body is a proper base for a sound mind.

The Publisher

AUTHOR`S PREFACE:

WHAT will Fertilizing with STONE-DUST ACCOMPLISH?

It will:
1. Turn stones into bread and make barren regions fruitful.
2. Feed the hungry.
3. Cause healthy cereals and provender to be harvested and thus prevent epidemics among men and diseases among animals.
4. Make agriculture again profitable and save great sums of money which are now expended either for fertilizers that in part are injurious and in part useless.
5. Turn the unemployed to country life by revealing the inexhaustible nutritive forces which, hitherto unrecognized are stored up in the rocks, the air and the water.

This it will accomplish.

May this little book be intelligible enough that men, who seem on the point of becoming beasts of prey, may cease their war of all against all and instead unite in the common conquest of the stones.

May mankind, instead of hunting for gold, racing for fame, or wasting productive forces in useless labors, choose the better part: The peaceable emulation in the discovery and direction of the natural forces for evolving nutritive products and the peaceable enjoyment of the fruits which the earth is able to produce in abundance for all. May man use his divine heritage of reason to attain true happiness by discovering the sources whence all earthly blessings flow and thus put an end to self-seeking and greed, to the increasing difficulties of making a living, the anxieties for the daily bread, to distress and crime - such is the aim of this little work, and in this may God aid us!

JULIUS HENSEL, Hermsdorf u. Kynast, October 1st, 1893

THE CAUSE OF THE DECADENCE OF AGRICULTURE:

The yield of the ground is steadily decreasing. Everywhere is distress. Our fields do not yield sufficiently abundant crops to compete with the cheap lands of the far West. To change this condition is the object of this book.

It is now 400 years since the second half of the world was discovered, but the whole earth is only now discovered, so far as the knowledge is concerned, of how the inexhaustible treasures may be utilized which are at our disposal in the nourishing forces of the rocks of the mountains. Instead of working this colossal mine men have bought the material for restoring the fertility of the exhausted soil in the form of medicine; i. e., chemical fertilizers.

For the last fifty years a dogma has crept into agriculture which calls itself "The Law of Minimum," namely:

"*That one* of the substances which the plant requires and which is contained in the minimum quantity in your fields you must furnish to it in the form of a fertilizer."

This false precept owes its reception solely to the defective method of chemical investigation which prevailed fifty years ago.

As there was found a considerable quantity of phosphoric acid and of potash in the ashes of all seeds, and as these do not exist in the air and must therefore be furnished y by the soil, it was very natural that the inquiry was started, how much of these substances necessary for the raising of plants is still at hand in the soil?

While the soil was then investigated and was treated with *muriatic acid*, in order that the substances contained might be dissolved, there were found only inconsiderable quantities of potash and of phosphoric acid in this solution, because the alkalies in the soil which are combined with silicic acid are as little dissolved by muriatic acid as, e. g., powdered glass. In order to be able to define the amount of potash, it is necessary first to drive out the silicic acid by the use of fluoric acid after having converted it into volatile fluoride of silicium; this method was not used by the former agricultural chemists. As in consequence thereof they overlooked the presence of potash, so also did they fail to notice the phosphoric acid which is combined with alumina and iron in the silicates, because when the iron was precipitated from the solution the whole of the alumina and phosphoric acid was precipitated with it; the further examination of the fluid solution therefore gave a negative result with respect to phosphoric acid, and this is also the case at this day if we work according to the old method.

The teachers of agriculture therefore announced:

"Of potash and of phosphoric acid, these most important nutriments of plants, there is only a *minimum* left in the soil; therefore, we must first of all supply potash and phosphoric acid to our fields."

To these two substances *nitrogen* was also added. Nitrogen in the form of vegetable albumen is on the average contained in such quantities in plants that its weight frequently exceeds that of the fixed constituents of the ashes. The following may serve to explain this: The affinity of the earthy substances (lime, magnesia and oxide of iron) and of the fixed alkalies with respect to hydro-carbons is quite limited; its sphere of operation is limited to eighteen molecules of hydro-carbons, as may be seen in the soaps, which consist of combinations of potash or soda with oleic acid ($C_{18}H_{34}O_2$,) or with stearic acid ($C_{18}H_{36}O_2$). Of like affinity with these earths and the fixed alkalies is the volatile alkali Ammonia N.H.H.H. This explains why when there are not sufficient earths carried up in the juice to complete the upbuilding of plants in their stalks and leaves their place is filled by ammonia, which, as before said, is formed from the nitrogen and the watery vapor of the air.

The wood in the trunk of trees contains no nitrogen at all, but the leaves of trees contain a quantity of nitrogen; the parenchyma of the leaves condenses it from the air because the sphere of action of the earths, which extends even unto the veins of the leaves, does not reach the parenchyma.

Now, in view of the great quantity of nitrogen found in the produce of the fields and of which agriculturists presuppose that it is derived through the roots of plants from the earth, they came to the same result as with respect to potash and phosphoric acid; i. e., they found only a vanishing „*minimum*" of it in the soil, and therefore they concluded: " Our crops have already consumed all the potash, all the phosphoric acid, and all the nitrogen; these substances are, therefore, in "minimum" proportions in the soil. If we are not to miserably starve we must bring this minimum in abundance into our fields in the form of manures."

The result is that the use of Super-Phosphates, *Sulphate* of *Ammonia*, *Guano* and *Chili-nitre* has enormously increased, but agriculture has entered into the sign of the cancer (retrogression), for it may easily be seen that if the cost of fertilizers amounts to more than the harvest the farmers must emigrate.

It took a long while before the teachers of agricultural economy, having the fact pointed out to them by practical farmers who judged with clear eyes and sound reason that crops of peas and beans rich in nitrogen prosper on soil entirely void of nitrogen, at least granted that leguminous plants derive their whole supply of nitrogen exclusively from the air, which as to full four-fifth consists of nitrogen. It is difficult for them to admit that other plants also do this, because their reputation and their income is mainly derived from the theory of potash, nitrogen and phosphoric acid. They explain this by asserting:

"There are *producers* of nitrogen and there *consumers* of nitrogen."

It is of course true that plants also assimilate such nitrogen as their roots find in the soil, but that is by no means necessary. The forest trees furnish us with a most convincing proof of this. Birches, beeches and oaks grow to gigantic size on bare rocks of granite and porphyry. To be convinced of this let anyone ascend the Herd mountains! Now as beech leaves and oak leaves contain one full per cent. of their weight of nitrogen, while beech wood and oak wood are devoid of nitrogen, the nitrogen of the leaves has evidently been furnished, not by the rock, but by the air.

It is manifest that if the soil were the proper source of nitrogen the roots being in immediate contact with the soil ought to show at least as much nitrogen as the parts above ground which are surrounded with air; but, on the contrary, they contain less.

For example, one pound of potatoes contains about 25 grains of nitrogen, but the green potato stems and leaves contain more than 42 grains to the pound, and it is from the plant that the tubers draw their nitrogen and not the reverse; for the potato herb, which in the beginning was so exuberant in juice, about the time the tubers mature becomes thin, hollow and light, because the juice containing the nitrogen descends into the tubers. So also does one pound of the green plant of the carrot contain about 35 grains of nitrogen, but the carrot-roots contain only about 14 grains to the pound.

We may also mention that just as the nitrogen descends into tubers it also passes up into the seeds, so that cereals show as much as 140 grains of nitrogen to the pound. The green stalk of grains show a similar proportion of nitrogen, while in a pound of straw there are only found 33 to 49 grains of nitrogen.

That the chemical fertilizers that are still all the fashion are a mere waste may be mathematically demonstrated by taking any example at random. I will choose for this the *sugar-beet* and the *carrot*.

The sugar-beet according to Wolff's tables shows the following ashes per kilogram (2,20 lbs.):

Potash:	3.8	Phosphoric acid:	0.9
Soda:	0.6	Sulphuric acid:	0.3
Lime:	0.4	Silicic acid:	0.2
Magnesia:	0.6	Muriatic acid:	0.3

According to atomic equivalents this would make 142 for phosphoric acid, 80 for sulphuric acid, 60 for silicic acid, 73 for muriatic acid, 94 for potash, 62 for soda, 56 for lime and 40 for magnesia.

Now the above
0.9 of phosphoric acid would saturate 0.6 potash.
0.3 of sulphuric acid   „   „   0.35   „
0.2 of silicic acid   „   „   0.30   „
0.3 of muriatic acid   „   „   0.40   „

Thus all the acids together   „   1.65   „

There remains, therefore, the following excess of bases:

Potash:	2.15
Soda:	0.60
Lime:	0.40
Magnesia:	0.60

Or if we count the 0.6 of soda, 0.4 of lime and 0.6 of magnesia equivalent with 1.65 of potash, then the entire quantity of potash in the sugar-beet, amounting to 3.85, would be at our disposal. This potash we may consider as combined with carbonic acid in the ashes, while it exists in the sugar-beet in combination with sugar, cellular tissue and albumen. Besides these 3.8 potash, 1.6 of nitrogen, or, in round numbers, 1.9 of ammonia is to be taken into account as being also an unsaturated basic constituent of the beet-root. From this it proximately follows: That the 3.8 potash cannot result from the manuring with sulphate of potash, for else it would need the presence of 3.25 of sulphuric acid, while there is only 0.3 present, nor can the 1.9 of ammonia be due to the sulphate of ammonia, else they would call for 5.0 of sulphuric acid instead of only 0.3. If, therefore, we manure sugar-beets with sulphate of potash and sulphate of ammonia these substances are to be regarded, as already stated, as largely wasted. As the source of the potash and the soda for the sugar-beets we can only consider the feld-spar, which, thanks to God, is still contained to a certain degree in the soil, while the nitrogen is furnished by the atmosphere.

The feld-spar in the soil will in the end of course become exhausted, and must then be supplied by manuring with the rock fertilizer.

A computation shows that to supply 0.3 sulphuric acid 0.6 gypsum which is combined with water will suffice; thus if the acre of land is to furnish two cwt. of beets it would need among other things only 13,25 lbs. Of gypsum.

As a parallel we will now consider the carrot. The ashy constituents of one kilogram (2.206 lbs.) are according to Wolff´s tables as follows:

Potash:	3.0	Phosphoric acid:	1.1
Soda:	1.7	Sulphuric acid:	0.5
Lime:	0.9	Silicic acid:	0.2
Magnesia:	0.4	Muriatic acid:	0.4

A comparison with the sugar-beet roots shows that the carrot contains somewhat less potash and magnesia but somewhat more soda and lime; besides this, the carrot contains about one-third more of phosphoric, sulphuric acid and muriatic acid. These variations seem to be caused by manuring with liquid stable manure; as to the rest we recognize that for the basic constituents of potash, soda, lime and magnesia in the carrots the pulverized primary rocks of the soil are the natural source.

We find that all plants, as also all animal bodies (for these are built up from vegetable substances), after combustion, leave behind ashes which always consist of the same substances, although the proportions of admixture vary with the different kinds of plants. We always find in these ashes potash, soda, lime, magnesia, iron and manganese combined with carbonic, phosphoric, sulphuric, muriatic, fluoric and silicic acids. These ashy constituents give their form and connection to the bodies of plants and animals according to the manner indicated above.

Now, inasmuch as the plants spring from the soil, it is manifest that the enumerated earthy or ashy constituents must be furnished by the soil. And, as in the soil these substances are present in combination with silica and alumina, the origin of the soil thence becomes manifest. It has arisen from the disintegrated primary rocks, all of which contain more or less potash, soda, lime, magnesia, manganese and iron, besides phosphoric, sulphuric, chlorine, fluorine, silica and alumina. From such earthy material from primary rocks, which have been associated with sediments of gypsum and lime, in combination with water and the atmosphere under the influence of the warmth and light of the sun, the plants which nourish man and beast originate.

Now, as all the enumerated earthy materials with the exception of silica and alumina enter into the crops that are taken away from field, it is clear that they must be replaced. If we desire normal and *healthy* crops, and that men and animals living on them should find in them all that is necessary for their bodily sustenance (phosphate and fluorate of lime and magnesia for the formation of the bones and teeth, potash, iron and manganese for the muscles, chloride of sodium for the serum of the blood, sulphur for the albumen of the blood, hydro-carbone for the nerve-fat), it will not suffice to merely restore the *potash, phosphoric acid* and *nitrogen*. Other things are imperatively demanded.

With regard to this I shall adduce one instructive example: The proprietor of an extensive estate wrote to me that he formerly manured with ammonia, super-phosphate and Chili-nitre, and although there was a steady retrogression in the yields, yet he continued to earn something. Of late, however, when he had passed over to manuring with iron slag and Chili-nitre, with a steady retrogression, at least neither rye, nor barley, nor oats would prosper, only, strange to say, wheat gave a tolerable yield. How could I explain this to him? To this question I gave the desired answer by pointing to the *ashy constituents*. The ashes of barley and oats contain five times as much sulphuric acid as wheat. The latter could still find its small requisite of sulphuric acid in the soil, but for oats, barley and rye these feeble remains did not suffice.

Now as we have seen that the primary rocks in the mountain ranges, porphyry, granite and gneiss, through the mellowing and crumbling influence of thousands of years (for nothing else is meant by „*disintegration*"), has produced the fertile soil which furnishes us with healthy nourishing plants, it may easily be seen that when such a soil has been almost exhausted of the elements that nourish plants through a cultivation of several hundred of years and a yearly turning over with the plough or the spade, the original natural strength cannot be restored to it by means of medicines and single chemicals, but this can only be effected by supplying new soil out of which nothing has grown, and the strength of which is therefore intact.

To gain such new soil we need not wait a thousand years till wintry cold, snow and rain crumble the rocky material and bring it down into the valleys. We have only to put our hands to work, and from the proper rocks obtain the necessary materials to rejuvenate the old and worn out soil and restore it again to virgin fertility.

HEALTHY AND UNHEALTHY PRODUCE

According to the chemical examination of the ashes which remain when plants are incinerated the average result shows about as much *potash* and *soda* as *lime* and *magnesia*: *silicic acid* somewhat - more than one-fifth of the sum of these four bases; *chlorine* about one-twentieth of the whole; *phosphoric acid*, one-sixth of the whole; but *sulphuric acid* only one-fourth in weight of the *phosphoric acid*.

Now, as granite rocks contain on an average 6% of potash and soda, while their contents of phosphoric acid are more than one per cent, granite by itself will ready fulfill the demands for vegetable growth, as may be confirmed by a report in the papers received while writing this. We read:

"In Deutmansdorf, Kneis Loewenberg, in Silesia, were found on the heap of refuse from the quarries there stalks of rye, with ears containing ninety to one hundred grains" (*General Anzeiger für Schlesien and Posen*, October 1, 1893).

As to chlorine, this mostly reaches our cultivated plants through manuring with liquid manure containing salt, and has been proved directly injurious to the growth and quality of many plants; in this respect it is sufficient to point to the evil effects of manuring tobacco with liquid manure. Chlorine is not found in wheat, rye, barley and oats, millet and buckwheat, linseed, apples and pears. plums and gooseberries, acorns and chestnuts, nor in the wood of any forest trees. We need therefore, not consider chlorine in fertilizing our fields.

Now, when I state that the given proportions of the ashes have yielded this average in comparing more than eighty analysis of the ashes of the various parts of plants, it need not be concluded from this that any particular plant, or the particular part of a plant, needs a quite definite proportion of ashy constituents; but it is found on the contrary that the earthy constituents of the same kind of plants differ in various ways. This explains why we find the same species of plants to flourish now on calcareous soil, now on soil formed from granite, gneiss or porphyry, as an example of which I shall only mention sheep's yarrow (Achilles millennium). This is effected in great part by the fact that potash and soda are interchangeable, but these two alkalies may also be replaced in most plants to a considerable part by the alkaline earths, lime and magnesia; but, of course, the nutritive value of the plants and the other qualities cannot then remain the same. Potash and soda may even be *wholly* lacking in a plant and they may be *entirely* replaced by lime and magnesia. As this fact is not as yet found in any book, I can not refuse a challenge as to proof. I name as my witnesses the royal mason, *Wimmel*, of Berlin, and engineer *Klug*, of Landshut. In company with these gentlemen i visited, on June 25th, of this year (1893), the loftily situated marble quarry near Rothenzechan. In the neighborhood of this marble quarry the vegetation is always somewhat behind in time to that of the valley, and so we found the dandelions still wearing their downy crown, while in the valley, by the end of May, they have passed away. We found such dandelions there growing immediately on the marble rock, where this had water flowing over it, and the flower-stalks reached the height of about a foot and a half. There was not, indeed, any great wealth of leaves, and the thick and high flower-stalks themselves could be broken like glass into pieces, which I did not weary in repeating before the eyes of my companions. Now this Silesian marble is a very white dolomite, consisting, therefore, of carbonate of lime and carbonate of magnesia; but it must also surely contain besides this some phosphate and sulphate of lime besides a trace of carbonate of protoxide of iron the presence of which is demonstrated in the moist clefts of the marble by a brown oxidation. These plants, therefore, grew on a sub-stratum of almost pure lime and magnesia. This extreme example convinces us that the alkaline earths (lime and magnesia) may really replace the alkalies (potash and soda) in the building up of plants, and this also furnishes us with the explanation why the iron-slag, as a pre-eminently calcareous fertilizer, unmistakably caused an increase of crops on fields which were deficient in lime. The same result might, indeed, have been reached more cheaply by directly spreading the lime on them. But there is another "But" in this matter, for in harvests it is not merely the quantity but much more the quality which has to be considered.

Even if the striking example cited makes it manifest that lime may in great part replace the alkalies in building up plants, giving to them the same *form*, and, indeed, making them of imposing size, nevertheless the *quality* and the *internal worth* of the products of the soil is considerably influenced by the difference in its basic constituents. I, therefore, mentioned, not unintentionally, that the flower stalks of the dandelion grown on marble could be broken like glass into separate pieces, while on the contrary dandelion stalks growing on soil containing *potash* may be bent into rings and formed into chains, as is frequently done by children. Potash makes pliable and soft, lime makes hard and brittle. *Flax* is a very good example of this.

Silesian linen made of the flax growing on our granite soil rich in potash is celebrated for its suppleness, softness and durability, while the Spanish and French linen produced on calcareous soils is hard, of little strength of fibre and of small value. What avails it then that the Spanish flax exceeds the Silesian by twice its length?

As with textile plants so with plants serving for nourishment and for fodder. It is manifest that where calcareous plants have not the same nutritive value as those in which alkalies and earth are harmoniously associated that the former are not as healthy as the latter. With reference to this, Dr. Stamm, who practiced in Zurich, (where in 1884 i saw a whole mountain of lime dug away), states that he nowhere saw so many examples of ossification of the arteries as on the Swiss soil so rich in lime; the fact that the drinking water is correspondingly rich in lime may contribute to it. The strong, bony frame of the Swiss strikes every one, even those travelers who visit Switzerland only for a short trip. This was an essential reason why Winkelried in 1386 at Sempach could with his strong-boned arms hold a whole dozen of lances of the knights, and 1400 Swiss could win the victory over 6000 Austrians who were fed on meat, wine and flour, and this, despite of their 4000 horsemen in armor.

How much influence nutrition exercises on temperament and breed may be seen from the breeders of fine horses. As Prof. Marossy communicated to me, Englishmen import the oats for their race horses from Hungary. Why? Because the granite of the Carpathian mountains is rich in potash, but contains but little lime. Potash makes supple, but lime makes tough and awkward. The counterpart of the world-renowned Hungarian saddle horses and carriage horses is found in the strong-boned Norman breed horse which derives its peculiarities from the French chalky soil, and could not be easily replaced as draught animals before the heavy stone wagons, the baggage wagons and the brewers wagons with their heavy loads of beer barrels.

And is it possible that the human race should be uninfluenced by its nutrition?

Let us make some comparisons: Wine contains almost only phosphate of potash, for the calcareous ingredients are precipitated during the fermentation as tartar. Hence the French *esprit*, the Austrian good nature, the artistic inspiration of the wine-drinking Italians. But like a stonewall in the battle stand the Pommeranian potato-eating grenadiers. In the ashes remaining from potatoes we find the following proportions: 44 potash, 4 soda, 64 lime, 33 magnesia, 16 phosphoric acid and 13 sulphuric acid. Sulphur is indispensable in the formation of normal bile and of tendons. Also hair and wool require much sulphur, about 5 per cent. of their weight.

After some such hints as to nutrition it cannot be indifferent what kind of crops we raise for our nourishment and with what substances our fields are fertilized. It cannot be all sufficient that great quantities are harvested, but the great quantity must also be of *good quality*. It is indisputable that by merely fertilizing with marl; i. e., with carbonate of lime, such a large yield may be gained as to make a man inclined to always content himself with marl, but with such a one-sided fertilization slowly but surely evil effects of various kinds will develop; these have given rise to the axiom of experience: "Manuring with lime makes rich fathers but poor sons."

Despite such experience, however, after a certain time, when those who experienced the damage have passed away, manuring with lime always again becomes a fashion. So even now. The harvests after manuring with lime are so favorable that there are those who expect their salvation from fertilizing with lime. Not long ago the German Agricultural Society granted a prize to a paper on "Fertilizing with Lime." But such prizes do not prove anything. Also a paper on Chili-nitre as a fertilizer received a prize. But how has this substance, so poisonous for plants and animals, fallen into disgrace!

Lime, indeed, is not directly injurious to plant growth, on the contrary it is useful and necessary, but everything has its measure and its limits. Lime can only produce wholesome cereals, vegetables and forage when there is at the same time a sufficiency of potash and soda.

"Too much of a thing is good for nothing!" In this connection I have to add a few things more. In the same way as lime and magnesia can replace potash and soda in the structure of plants so all these four constituents can in great part be replaced by basic *Ammonia*, without any resultant appreciable change in the *form* of the plants, except that this appears strikingly luxuriant and *rich in leaves*, as the milfoil on and near the mounds of cemeteries.

Such a substitution of ammonia for the alkalies and the alkaline earths corresponds in some degree to the relation between potash-alum and ammonia-alum, which are so similar in form that they cannot be distinguished without a chemical examination. In an analogous manner the muriate of ammonia has quite a similar taste to the muriate of soda, and the sulphate of ammonia almost the same bitter taste as sulphate of soda (Glauber's salt) and sulphate of magnesia (Epsom salts), but the *effects* of these salts vary considerably.

A particularly interesting example of the fact that the appearance of plants in which ammonia has largely taken the place of the fixed alkalies and earths is found in *tobacco leaves*. Only specialists can at once recognize their quality at a glance; the great majority only perceive the difference when the leaves, made into cigars, are lighted. Then the one kind, grown on the Virginia soil, rich in magnesia and lime, gives us light, loose ashes and a fine aroma, while the product of Vierraden (Prussia), manured with stable dung and liquid manure, in which ammonia takes the place of lime and magnesia, "coals," and diffuses an unpleasant odor. It is quite similar with plants raised for nourishment or for fodder. The inability of offering resistance, as seen in the "lodging" of grain after manuring with dung and liquid manure, after a long rain, and, in accordance with this, the grain also which is harvested from such a field has no firmness; it becomes soft in grinding, smearing the mill-Stones, so that no grain raised with stable manure can be ground without mixing it with Western or California grain, and it has always a lesser value. So the barley raised with stable manure produces a malt which the brewer refuses to buy, as it would spoil his beer.

Now, as these ammonical plants lack the internal firmness and the ability of offering resistance, so also they cannot be healthy for animals when used as fodder, for the animal bodies have no consistence without earths. But these earths are subject to elimination owing to respiration. The ashy constituents of the blood-disks, which are oxidized by respiration; i.e.; sulphate and phosphate of lime, magnesia and iron, are eliminated from the organism with the secretion of the kidneys, as also the bases present in the flesh of the muscles; i. e., potash and soda; for the muscular substance also is oxidized through the oxygen of the arterial blood.

Now, as the earthy or ashy constituents, which are specifically necessary for the albumen of the blood, as well as for the flesh of the muscles and for the renewal of the bones (for *all* parts of the body are subject to this mutation of substance), are not replaced by the substances in the fodder, it is an unavoidable sequence that a relaxation and loosening of the tissues, brittleness of the bones and every kind of disturbance of health should take place with our cattle. 1 shall only adduce one single very instructive example in proof of this out of my neighborhood.

The hotel keeper in Carlsthal, near Schreiberhau, in the Riesengebirge, kept twelve beeves. The manure from the cattle he conveyed to a swampy meadow, which up to that time had only produced sour grasses; but after the stable manure was applied it yielded so luxurious a growth of grass that he used the abundance to feed his twelve oxen and cows. It was not long, however, before the cattle became decrepit and ten of them died. The cause of this was the fodder grown from stable manure, in which ammonia supplied the place of the fixed alkalies, potash, soda, lime and magnesia. The other two beeves were quickly sold, for they suffered from lickerishness; i. e., they refused their food and gnawed instead the cribs and other wood in the stable. For all wood contains about three per cent of earthy substance, and the cattle craved these earthy substances in order to gain firm flesh and bones. The two oxes recovered when their new owner gave them a different fodder.

This same reason serves to explain other cases lately observed. It has been found that some kinds of pork do not bear pickling. While salt and nitrate of potash insure the keeping of pickled meat, the meat of certain hogs when lying in brine very soon passes into putrefaction, but into a decomposition different from the usual kind. The process that takes place is like what is called the "cheesy degeneration", which chemically means that the connective and muscular tissues decompose into peptones (Leucin and Tyrosin) as during digestion.

To explain this phenomenon we must consider the cheesy degeneration of the lung tissue in consumptives. In their blood there is also always a deficit in lime and sulphur, which are absolutely necessary in the formation of red blood-disks.

Now, on inquiring why this pork when pickled underwent such a peculiar change, it was found that the animals had been fattened with Fray Bentos Meat Powder. But the lean meat contains at its chief ashy constituent only phosphate of potash with almost imperceptible traces of lime and sulphur. Lime, indeed, is not found in the meat but in the bones, which are devoured by the tiger and the dog, but not by man. Therefore we have to gain the calcareous supply for our blood, our bones and our teeth from calcareous corn and from vegetables rich in lime. As our present fine flour, freed from bran, is furnished us almost entirely devoid of sulphur and lime we need not wonder at the great number of modern maladies.

Now when hogs are fed with Fray Bentos Meat Powder, devoid of lime in place of vegetable food, rich in lime, they cannot acquire,a strong bony frame, and in consequence we need not wonder at the flaccidity sponginess and easy putrefaction of the meat of such animals. If they had not been slaughtered in good time these helpless animals likely would have succumbed to some hog disease.

From this we may draw our conclusions with respect to human health. Many a one considers a meat diet to be a godsend, but is plagued on that account with rheumatism, asthma and corpulency, to cure which he is ordered to drink some mineral waters which contain lime, magnesia and sulphate of soda.

To return to agriculture and the feeding of cattle. Nitrogenous foods are supposed to be strength-givers; this is a theoretic error full of fatal consequences for agriculture. We never have had as many cattle plagues as we have had since artificial fertilizers and "strong" foods have been "en vogue".

The theorists in nutrition who demand that man should have so much of hydrocarbon, so much fat and so much albumen have evidently little conception of the close relationship in which these substances stand to one another, by which the one may pass over into the other; e. g., the hydro-carbon sugar through the adjunction of earths and ammonia becomes albumen. But albumen easily undergoes a change into fat, as may be seen in cheese, and also from the manner in which the meat of the ham passes into fat. The same transformation takes place in nutriments containing hydro-carbons; e. g., the malt sugar of beer drinkers and the starch of grasses. Many an ox accumulates a few hundred points of tallow and yet is not fed with fat or butter, but with grass, hay and grain.

The so-called "strong" food of cattle, therefore, amounts to nothing and ought rather to be called poison-food. The truly strong food for cattle consists of mountain herbs, rich in earth, when these besides alkalies also contain lime and magnesia. Just think of the dairy cows of the Swiss Alps and of the cattle in Holstein that derive their strength from the grass of the marshes which are not fertilized with stable manure, but which are preserved in their lasting fertility by the neighboring rocky highlands, which continually enables the rain to wash down the soluble rocky material which enriches the meadows.

As a counterpart to the pork raised from Fray Bentos Meat Flour I will mention here an example from my own observation. Here (on the Kynast) I kept two sheep, I once saw them eating lime from the wall of the stable, as chickens do when they need lime fer their egg-shells. From this I concluded that the grasses growing on my soil, in which there is little lime, did not supply them with sufficient sustenance for their bones. I, therefore, mixed some whiting with their cooked roots and this craving for lime ceased. When I at last sold the animals to the butcher he was so much pleased at their solid meat that he desired to bespeak immediately some sheep for the next year.

I would here mention that the sheep raiser, Mr. von Wiedebach, in Guben, inquired of me whether the principles of my "Makrobiotik" were not applicable to cattle-raising, especially so as to put an end to the mortality among sheep and to the mouth-disease and the foot-rot; I advised him to give in certain proportion precipitated chalk or whiting, flowers of sulphur and copperas, as a periodical addition to their food; and he has repeatedly assured me that by these means he has in very many places, whither he has been called as a specialist, put an end to the mortality of the cattle, and has brought them in other places to a normal state of health.

Chemistry teaches us that the characteristic nature of the albumen rich in ammonia consists in the easy interchangeableness of their atomic groups; but just on this account muscular fibre and connective tissue can be built up from the albumen of the blood, but every case has two sides. The ease with which the constituents of albumen can be shifted also favors their chemical decomposition. Need I to mention here the savory taste of fresh-laid eggs and the smell of rotten ones? Intelligent people have long since perceived that feeding with albumen does not do all that the theorists claim. It does no pay for its expenses.

The chemical "strong" food for the *soil* in shape of the Chili-nitre, which contains nitrogen, and has been awarded the premium over all its competitors, has proved a miserable failure; but the theorists are indefatigable. They now advocate chemical "strong" food for *cattle*, and there are many people who put these latest theory into disastrous practice.

All of us have to bear the evil consequences of this. Does not stable-feeding cohere with this "strong" feeding and this forced fattening? And does not the stable air so poor in oxygen cause the murrain of cattle? And does not the mortality of our children spring from the cow-milk poor in earths? That in consequence of fertilizing with stable manure crops poor in earths are produced is indubitable after what has been stated above. From these nutriments poor in earths again follows a host of ills: nervous debility, nervous sufferings, decomposition of lymph and serum, which are continually becoming more prevalent. Among these diseases are anemia, chlorosis, scrofula, swelling of the lymphatic glands, cutaneous diseases, asthma, catarrhal states, nervousness, epilepsy, gout, rheumatism, corpulency, dropsy, consumption, diabetes, etc., as I have demonstrated anatomically and physiologically in a manner easily comprehended in the book "Makrobiotik" or "*Our Diseases and Our Remedies*". Fertilizing with stone meal will in future give us normal and healthy crops and fodder.

WHAT SHALL WE DO WITH STABLE MANURE?

So long as attention was not called to the fact that new earth from pulverized primitive rock, together with the carbonate and the sulphate of lime, forms the best and most natural fertilizer for an exhausted soil, men directed their attention to that part of the food which cattle do not assimilate, but excrete, for manure. So men came pretty generally to the position that we must bring dung to an exhausted field, else nothing can grow. Now in order to get manure, we must raise cattle; for these stables and and attendants are necessary, and a considerable area of land must be devoted to provide the necessary fodder. Now since it is said that without manure nothing can grow, manure must be used to make fodder grow on which cattle feed in order to produce manure for more fodder. In such a circle of life where does the advantage of keeping cattle come in? The raising of cattle only pays in mountainous regions, where the fructifying dew transforms the stones into herbs, or in the marshes irrigated by canals, for here the subsoil is naturally moist, and without water nothing can grow. In marshy regions the raiser of cattle can put his hands in his pockets and look on while the ox "eats money into his pocket"; but elsewhere the ox rather eats money out of the owners pockets than into them.

But anyways the production of milk, butter, cheese and wool, as well as the necessity of having horses for driving, makes the raising of cattle and horses a factor that must be taken into consideration.

Now as all cattle produce manure, solid and liquid, the question arises: "What shall we do with it?"

The fact that stable manure undoubtedly promotes the growth of plants gives to it a certain value. This value does not depend upon the nitrogen but on the earthy or ashy constituents which it contains, and on the combinations of hydro-carbons; i. e., these carbonate carbohydrates do not need to be first produced by the sun, but may be utilized by a simple change of grouping and may be compared to ready-made building stone that may at once be built into the plants with this result that their growth may take place in the cool springtime *more quickly* than where the warm sun must do the whole work of drawing the carbon out of the carbonate rocks in conjunction with water. Still this advantage will no more be considered so decisive because the same result, yes, a result almost four times as great according to my experience, may be attained by a judicious mixture of rocks in a finely powdered state. This stone-meal is dry while manure is moist, and for that reason the former is worth at least four times as much because more condensed, while in addition thereto the earthy constituents have mostly been eliminated out of the latter by passing through the animal or human body, and the stone-meal mixture contains these in abundance. But of course not all earthy constituents have been taken out of the dung, for of some a superabundance may have been provided, part of which is still contained therein.

Such manure, therefore, is by no means without value; as animal bodies contain about 80% of water so there are also considerable quantities of water contained in crops. Dry hay, e. g., will in the kiln still lose 15% of water; and green fodder and vegetables contain a full 75-80% of water; in some root crops the water amounts even to 90%. Considering its properties of water stable manure is not to be valued so highly, because only an equal weight of crops can be procured therefrom. Still even this is sufficient to keep us from rejecting it. Only it should be deprived of the injurious qualities clinging to it owing to its excessive quantity of nitrogen. As far as the manure is concerned, indeed, little damage is done, for despite those erudite in manures the simple farmer spreads the liquid manure over the fields, where the ammonia N2H6 is oxidized into nitrogen N2 and water H6O3 Before this process is completed, or at least before the ammonia has been very much diluted, as on the irrigated fields, nothing will grow from it. The most important point lies in this, that it is not the nitrogen which is combined organically with hydrocarbons as in leucin, tyrosin and hippurate of lime which is the most injurious factor of the dung, but the *carbonate of ammonia*, which is formed from the *urea* of the liquid manure. Free ammonia is a poison to plants.

Ammonia is not only poisonous for plant-roots, but is also poisonous for animals, producing paralysis, even when diluted in the blood to a mere trace. In this respect I shall report a case from actual life respecting stable manure; wherever a similar state prevails a lesson may be drawn from what I report.

In certain cavalry barracks there was a rule that in summer the bedding from the horse stables should be spread in the morning on the open place before the stables to dry, and then be used again in the evening. In the stables of these barracks a remarkable mortality of horses developed, and what was the cause? The liquid manure in the straw became more and more concentrated, and carbonate of ammonia in excessive quantity was thence generated, because urea in a moist state in transformed into this substance.

```
      HHN        H                  HHHN
(Urea CO    plus O   minus COO =   (Carbonate of Ammonia)
      HHN        H                  HHHN
```

Such ammoniacal vapors are indeed perceptible in every horse stable, but in those military stables this evil was extreme. In stepping near the cribs, the rising ammonia vapors irritating the mouth and nostrils caused catarrh, and the eyes would gather tears. Now the heads of the horses were bent down over the cribs, they continually inhaled concentrated fumes of ammonia. This acted in a paralyzing manner on the nervus vagus and its branches in the respiratory organs and in the abdominal system. The horses were seized with fever, stopped eating and died. The veterinary physician did not recognize the carbonate of ammonia as the real cause of the appalling numbers of cases of disease and death, but according to his dictum the stables were infected with *Bacilli*. A thorough disinfection with carbolic acid was therefore ordered. For this purpose, of course, the bedding also with all its "bacilli" was ordered out and the learned veterinary physician gained a brilliant scientific victory, for after throwing out and burning the bedding and whitewashing the walls the mortality ceased for the time being.

In my book, *"Das Leben" (Live)*, I have recommended the transformation of the carbonate of ammonia which comes from the liquid manure into odorless sulphate of ammonia and carbonate of lime by strewing the stables with gypsum. Thereby the solid and the liquid manure are freed from their injurious qualities, which are manifested wherever the manure is heaped up a foot deep before it is removed and fresh bedding substituted for it. Those who have hitherto given no heed to the warning fumes of carbonate of ammonia and its evil consequences may continue to consult the veterinarians how to put a stop to the prevalence of cattle diseases.

We have already shown how the carbonate of ammonia may be rendered harmless. Now, in order to largely increase the value of the manure, the primitive rocks containing potash and soda, reduced to powder, should be scattered over the fields before the manure is spread. Thereby the nitrogenous hydro-carbons of the solid and liquid stable manure are prevented from entering into a decomposing fermentation, which give rise to unwholesome ammoniacal products of decomposition, which in part through capillarity rise into the plants without being transformed into vegetable substance; such plants when cooked manifest an ill odor, as may be seen in vegetables raised on fields irrigated with sewerage. Of late even roses are said to be cultivated on such irrigated fields near Berlin, but the home of the Bulgarian rose, that yields the attar of roses, lies at the foot of the Balkan mountains, which consist of granite, gneiss and porphyry; i. e., the rose demands a soil of disintegrated primitive rocks, or with us it demands as a fertilizer pulverized rocks. All roses fertilized with sewerage are infected with leaf-lice Whoever undertakes rose culture on such fields need not expect success.

In order to lay down once more the value of stable manure, and of excrementitious matter in general, it is demonstrable that nitrogenous ammonia is *injurious*. What is effective consists of the combustible *hydrocarbons*, which are ready *building material*, and further of the earthy or *ashy constituents* to which the *hydrocarbons* cling; for the hydro-carbons by themselves are rather injurious than useful to the growth of plants. This may be seen when we pour petroleum on the soil of potted plants, but *hydrocarbons combined with bases and soluble in water* advance the formation of leaves. I summarize then as follows:

1. Nitrogen in the form of carbonate of ammonia is directly injurious to the growth of plants.

2. Nitrogen is *unnecessary* as a fertilizer for the growth of plants if the soil contains a sufficiency of fixed basic substances (alkalies and alkaline earths). The proof of this in afforded by the fruitful calcareous soil of the Jura, which is not manured with nitrogen, so also the illimitable pasture grounds in America as also the vegetation of our German mountains. If plants find at their disposal for their growth a sufficiency of fixed bases they receive an ample supply of the complementary nitrogen from the air, four-fifths of which consists of nitrogen.

3. The nitrogen of the solid and the liquid manure may be used in the construction of plants, but in order to produce crops useful to health it is necessary to add to it a sufficient quantity of alkalies and of alkaline earths in the form of stone-meal as a counterpoise. By so doing we not only preserve, but especially amend, the nature of stable manure.

WILL FERTILIZING WITE STONE-MEAL PAY?

Some people say: "With such nonsense as Hensel's stone-meal I shall never have anything to do; nothing *can* grow from it" or "Useless dirt". This is the cry of men who have no chemical knowledge, yet two hundred farmers in the Palatinate testified before court that fertilizing with stone-meal showed far better effects than those from the artificial manures used hitherto.

"What do you say to this?" asked the judge of the young man who had declared the stone-meal a swindle (being himself a dealer in artificial manures). "I don't say anything to it; the people deceive themselves," replied the young man, who was fined for a too libelous tongue.

Since then persons who traffic in artificial manures are good enough to allow: "We will not deny that Hensel's stone-meal may have a certain effect, but this is far too slow and too small; for the silicate bases are almost quite insoluble and it will have to disintegrate for many years. These people also are deficient in chemical knowledge.

The silicates have indeed little solubility in water and hydrochlorid acid, but they do not resist water and the forces of the sun.

Of course, in speaking of the solubility of silicic acid we must not compare this with the solubility of common salt or sugar. Lime would sooner do for comparison, for of this one part dissolves in 800 parts of water. Silicic acid is somewhat less soluble, for little more than one-half of a grain is dissolved in 1000 grains of water. All hot springs contain silicic acid in solution together with other substances from the primitive rocks.

Men who say that silicates of bases are insoluble are contradicted by the trees of the forest an well as by every single straw. Oak leaves, on combustion, leave 4,66% of ashes, and of these fully one-third consists of silicic acid. How can this come into the leaves unless the ascending sap conveyed it in solution?

The accumulation of the silicic acid in the leaves is the result of the evaporation of the water which conveyed it.

From the forest tree to the straw! In the ashes of the straw of winter wheat, two-thirds consist of silicic acid. In the beard of the barley the proportion is still greater. This yields nearly 12%, of ashes, and 8,5% of this consists of silicic acid.

Still more striking is the solubility of silicic acid in the stems and leaves of plants which grow in water or in wet soil. Reeds, e. g., on combustion, leave 3,33% of ashes, more than two-thirds of which is silicic acid.

Sedge or reed-grass yields 6% of ashes, of which one-third is silicic acid. That sedge is at the same time rich in potash proves in the most striking manner that it needs only irrigation to make silicate of potash available for plant growth. Shave-grass (Horse-Tail) leaves 20% of ashes, half of which consist of silicic acid. From this it may be seen that only in those parts of plants which rise above the water, so that evaporation can take place, silicic acid is accumulated. But in the water itself this very solubility of silicic acid stands in the way of its accumulation. The best proof of this we find in seaweed. This leaves behind it a greater proportion of ashes than most plants, namely, up to 14,6%, but only one-fiftieth of this is silicic acid. The remainder mainly consists of sulphate and muriate of potash, soda, lime and magnesia; these the seaweed concentrates and combines with its cellular tissue, for sea water has not 14,6% but only about 4% of saline constituents.

This is sufficient to prove that with respect to vegetation silicic acid and silicates are not insoluble; on the contrary, they enter, like all other saline combinations, into the most intimate combination with glycolic acid, C 0 0 C H H, which is intramolecularly present in the cellulose of plants, so also with the ammonia of the chlorophyl so that the silicates cohere with the plants growing from them as an organic whole. We may easily convince ourselves of this on tearing a weed out by its roots. Then it is seen that the root-fibres of most of the plants are everywhere entwined around little stones which, dangling, still cleave to them and can only be torn away from them by violence and by tearing off some of the fibres.

So the objection as to the insolubility of silicic acid is invalid both theoretically and practically.

In reality we cannot find a root, a stem, a leaf or a fruit which does not contain silicic acid. This fact must be known to every teacher of agriculture. How then can these teachers deny the solubility of salicic acid in vegetation, as many of them who advocate artificial fertilizers do?

The men interested in artificial manures who thought that they had attended the funeral of stone-meal as a fertilizer have learned nothing from history, or have at least forgotten that *every* new truth has first to be killed and *buried* before it can celebrate its resurrection. Besides I do not stand as isolated as these people suppose, for I have the light of truth and of knowledge on my side:

> *"He who fights for truth and right*
> *E'en alone, has strength and might"*

I can also call to my aid a whole army of men who understand something of chemistry and of scientific farming, and their number, at this day where science is making such giant strides and hundreds of well edited agricultural papers are ready to support the interests of the farmer, is daily on the increase.

What is lacking at present is that the manufacture of stone-meal should be undertaken by men of scientific attainments who at the same time have sterling honesty, so as to make it certain that farmers will actually receive what is promised and what has proved itself so useful hitherto. I have received innumerable requests from farmers who asked for this mineral manure, but I had to answer them that with my advanced years I could not actively engage in this manufacture. The whole subject is of such immense importance for the common welfare that it is my wish to see this work placed into hands that are *thoroughly reliable*. I but point the way for the benefit of the human race.

The practical point to settle is how far fertilizing with stone meal pays, what yield it will afford, thus whether it will be profitable for the farmer to use it. I shall therefore treat this subject as exhaustively as possible and give an exact account of the results obtained.

It must here be premised that the *fineness of the stamping or grinding* and the *most complete intermixture* of the constituent parts are of the greatest importance for securing the greatest benefit of stone-meal fertilizing. A manufactured article of this kind has recently been submitted to me which showed in a sieve of moderate fineness three-fourths of the weight in coarse residuum. But as the solubility of the stone-meal, and thus its efficiency, increases in proportion with its fineness, the greatest possible circumspection is required in grinding it. The finer the stone dust the more energetically can the dissolving moisture of the soil and the oxygen and nitrogen of the air act upon it. A grain of stone dust of moderate fineness maybe reduced in a mortar of agate perhaps into twenty little particles, and then every little particle may be rendered accessible to the water and the air, and can, therefore, be used as plant food. Thence it follows that one single load of the very finest stone-meal will do as much as twenty loads of a coarser product, so that by reducing to the finest dust the cost for freight and carriage and the use of horse and cart would amount to only one-twentieth. Therefore we can afford to pay unhesitatingly a higher price for the finest stone-meal that has been passed through a sieve than for an article that may be not so much a fine powder but rather a kind of coarse sand.

The average contents of ash in cereals is about 3%. Thence, from three pounds of pure vegetable ashes we could raise a hundred pounds of crops. Now, as stone-meal properly made contains an abundance of plant food in assimilable form, it may be calculated to produce 4 cwt. (200 kilograms) of cereals, or that an annual use of 6 cwt. to the acre will produce 24 cwt. (1200 kilograms) of grain. On this basis every farmer can calculate whether it will pay. But in reality the harvest will be far greater, because even without the stone-meal most fields still possess some supply of mineral nutriment for plants which will become effective in addition thereto. Such being the case we need not consider the fact that not all the stone-meal is used up completely in the first year, but yields nourishment to plants even in the fifth year, as has been shown by experiments. That no mistake would be made by using double the quantity on an acre, or 12 cwt. (600 kilograms) instead of 6 (300 kilograms), is manifest, for the prospects of a still greater yield would thereby be improved, but in applying 12 cwt. (600 kilograms) an abundance would be supplied; even five or six times that quantity would be far from causing an injury to the soil, but we cannot force by excessive quantities of stone-meal a correspondingly higher yield of crops for the reason that within a definite area only a definite quantity of sunlight can display its activity, and on this factor the growth of the crop mainly depends. There is, therefore, no use in passing beyond a certain quantity of mineral manure; it could only come into use in subsequent years, and it appears to be more practical to supply the amount needed each year.

I will now present in summary form the quintessence of the significance of this natural fertilizer:

1. The point to be gained is not only a *greater quantity* of produce, but also a *better quality*. Sugar-beets gain thereby more *sugar*, this, according to experiments made, may amount to 75% more than hitherto. Potatoes and cereals show a greater proportion of *starch*. Oil crops (as poppies, rape, etc.) show more seedvessels and a corresponding increase in *oil*. Pulse, such as beans, peas, etc., yields -more *lecithin* (oil containing phosphate of ammonia as the chemical basis of nerve substance). Fruits and all vegetables receive a more delicate *flavor*. (The vegetables in my garden have become famous with our neighbors and our guests, so that they ask: "How do you manage that?") Meadows furnish grass and hay of more *nutritive* value. Vines form *stronger shoots*, give *sweeter grapes* and are *not touched* by *insects* and *fungous diseases*.

2. The soil is steadily built up and improved by this natural fertilizer, as it is progressively "*normalized*"; i. e., shows gathered together potash, soda, lime, magnesia, fluorine, and phosphoric and sulphuric acid, etc., in the most favorable combination. There is hardly *one* cultivated field which by nature is normal at the present time. Either lime prevails or we have a clayey soil, which through its excess of clay refuses to let the rain water pass, and by its toughness obstructs the access of the atmospheric nitrogen and of the carbonic acid; or we have a mere sandy soil (quartz), or again the soil has humus in excess, like the moor-land soil. This latter is characterized by a predominance of lime and magnesia on the one side, while sulphuric bases are two or three times in excess of phosphatic bases, as is shown by the analysis of the ashes of peat.

3. The value of the new fertilizer with respect to the wholesomeness of nutritive plants and fodder depends in great part on the careful and intimate comixture of its several constituents, so that with every little dust of potash and soda the other nutritive elements required to co-operate in the harmonic construction of plants are at their disposal in proximate vicinity. As a contrast to this, in a one-sided fertilizing with lime it may happen that the plant contents itself with the lime so that the other constituents of the soil are not drawn into co-operation for the growth of the produce, because they are not within the nearest proximity of the root-fibres. This is, of course, of great importance to the quality and the nutritive value of the plants.

4. For raising nutritive plants and fodder which may afford *wholesome nourishment* I deem it of the greatest importance that no substances should be used which lead to *ammoniacal decomposition*. By such additions we may indeed produce a luxuriant, excessive growth that blinds the eyes, and in which the abundant formation of leaves by means of nitrogen forms the chief part; but no healthy growth is effected thereby. From this point of view I would also deprecate the use of so-called fish guano. Every one knows how quickly fish pass over into putrefaction; there is formed at the same time a considerable quantity of *propylamin* ($C_3 H_6 NH_3$) which is an *ammoniacal base*. The manure manufactured in Sweden from fish-guano and powdered feldspar does not, therefore, merit the esteem that it claims.

A CHAPTER FOR CHEMISTS
The Chemical Process in the Growth of Plants which are the Basis of our Nutrition

> *"Every leaf in the hedge sings to the wise*
> *An excerpt from the wondrous lay of Creation!"*

The sum and substance of the growth of plants consists in creating out of burnt substances through the electrically decomposing forces of the sun material which may again be burned.

To take an example: A stearine candle, consisting of hydro-carbons (C H H), in a twenty-four fold aggregation, is consumed by means of the oxygen of the air into carbonic acid or carbon dioxide (C O O) and water (H H O) and these same products of combustion may through the vegetative processes in plants be again wholly or partially changed back into hydro-carbons. This is effected by separating from the carbonic acid dissolved in rain water or combined with the moisture in the soil, water and together with this oxidized water (peroxide of hydrogen). In this way there arise from two molecules of carbonic acid and two of water, first of all oxalic acid (C2 H2 O4) and peroxide of hydrogen (O H H O).

$$\begin{matrix} C\,O\,O, H\,H\,O \\ = \\ C\,O\,O, H\,H\,O \end{matrix} \quad \begin{matrix} C\,O\,O\,H \\ + \\ C\,O\,O\,H \end{matrix} \quad \begin{matrix} H\,O \\ \\ H\,O \end{matrix}$$

The peroxide of hydrogen passes into the atmosphere decomposed into watery vapor and oxygen, while oxalic acid arising as the first product of the reduction of carbonic acid caused by the action of the sun is found combined with lime in all vegetable cells. Formerly this *first process of growth* (for oxalic acid arises from the *accretion* of two atoms of hydrogen to two molecules of carbonic acid) was not at all understood. It is hardly four years ago that I heard a teacher of agricultural economy say: "Lime has no value in the growth of plants, it is rather injurious than beneficial. The plant knows not at all what to do with the lime; in order to rid itself more easily of it it takes it up as oxalate of lime in the cells".

Oxalic acid derives its name from the fact that chemists first discovered it in wood-sorrel (Oxalis) in the form of a combination of oxalic acid with lime. From the oxalic acid there proceeds in a continuous reduction *sugar, the material of plant-cells* and *starch*.

The sugar that has been produced from a symmetrical grouping of two molecules of hydro-carbons, two of carbonic acid and two of water:

$$\begin{array}{ccc} \text{H O H} & & \text{H O H} \\ \text{H C H} & = & \text{H C H} \\ \text{H C H} & & \text{H C H} \\ \text{O C O} & & \text{O C O} \end{array}$$

and which, therefore, is not yet a complete product of reduction, produces with the separation of carbonic acid and of water through a heaped up grouping together of hydro-carbons, which remain yet combined with a molecule of formic acid, C O O H H, (this second product of production or rather product of addition to Carbonic acid) then the vegetable *oils* (olive oil, almond oil, poppy oil, rapeseed oil, linseed oil, etc.).

Furthermore from sugar, which is exhibited in all young plants during their sprouting after receiving watery vapor and nitrogen from the air, and, indeed, after again separating peroxide of hydrogen, while ammonia arises, there are formed N2 H13 O6 NH6 H6 O6 the numerous kinds of *vegetable albumen*.

The simplest kind of vegetable albumen is found in asparagus, in the juice of asparagus:

```
O  HH  O  HH
C  CC  C  NNHH
O  HH  O  HH
```

a combination of ammonia with malic acid, which is a step towards the formation of sugar or rather a product of splitting off from sugar.

This asparagin is found not only in asparagus, but to select an example which may easily be demonstrated also in the young roots of thistles (which are weeded out from asparagus beds) and which taste very much like raw asparagus, and it is also found in the sprouts of very many other plants.

As the simplest of all kinds of vegetable albumen asparagus is the best exemplification of the fact that in albumen intramolecular *gelatine sugar* is contained:

```
H  O  H
C  C  NH
H  O  H
```

Of the latter, however, it is ascertained that on account of its contents of carbonic acid it can condense into an organic whole with itself *basic* substances (potash, soda, magnesia, oxide of iron and oxide of manganese), and owing to its basic ammonical substratum it also condenses *acids*, and accordingly also at the same time both bases and acids (*e.g.*, sulphate of magnesia, phosphate of lime, the silicates of potash and of soda, fluorate of lime), besides manganese and oxide of iron, and there arise indeed on account of the contents of the hydro-carbon (H C H) in the gelatine sugar from insoluble substances soluble combinations after the analogy of the insoluble sulphate of baryta and the ethyl sulphate of baryta which is soluble in water.

And so we may comprehend how from earthy elements in combination with sugar and nitrogen there can arise in endless modifications the most numerous varieties of vegetable albumen, according as the soil furnishes various substances.

In this the electrolytic force of the sun plays the part of the architect. Like as in the galvanic bath the atoms of the reduced metals apply themselves into a connected covering without a gap, so the solar forces cement together the reduced elements of the hydro-carbons with phosphates, sulphates, muriates, fluorates, silicates and carbonates of lime, of potash, soda, magnesia, and of the oxides of mangenese and iron together into the edifices which, as grasses, herbs, bushes and trees, refresh our eyes with their leaves and flowers, while their fruits serve to nourish man as well as the animal world.

But it is to be noticed that the above-mentioned processes only take place under the supposition that the carbonic acid, which lays the foundation out of which the hydrocarbons arise, find basic substances (potash, soda, lime, magnesia, etc.) with which they can condense themselves into firm combinations. Therefore the firm earth is the absolute condition for all vegetable growth, there is no vegetation in the air alone; nor must water be lacking (H H O), for its hydrogen (H H) being combustible in itself renders the groups of hydro-carbons combustible.

Now, as the process of our life represents nothing else than a continual combustion of our bodily substance by means of the oxygen respired, with the condition that to replace the substance consumed during the day by oxidation, during the night new combustible material must be supplied: by the contents rich in soil of the lymphatic vessels to the numerous nervous sheaths as the oil of life, and to the renewing blood new albuminous substance. Our life could not continue if we should not renew so much of the bodily material as is chemically consumed by the oxidizing respiration by means of the periodic supply of food. Every disturbance in the regular supply of food has the most manifest effects on the state of the soul. The inexorable demand for new material in place of the bodily substance which is breathed away makes even men, who by nature are kindly, angry and regardless of others when their food is kept back. And so cause and effect join themselves into a mischievous chain.

As the means for procuring food consists in by far the greater number of callings in coined money, and this is only given as a reward for work done, the question arises: What can the man do who has no opportunity and chance to find paying work? He will and must eat. If we can assist every one in getting a supply of food the main- spring for lying, deceit, stealing and numerous crimes vanishes.

Food is supplied to us in the first place by the immediate produce of the ground, and only in the second place by the fat, flesh and blood of domestic animals produced from grasses and herbs.

Now, as it is a primary chemical condition that earthy material, air, water and solar forces must be present in order that plants may grow, it is the all-mother Earth which surrounded by water and earth and fructified by sunshine, nourishes men and animals through the crops produced; and at same time it clothes animals, as their skin causes the hair to sprout forth, which contains sulphur and silica, and the hair isolating keeps together the bodily warmth and the bodily electricity.

Man, whose producing spirit desires occupation and to whom is granted the wonderful mechanism of the fingers, has the advantage that he can weave his garments according to the season, either of flax and cotton or of the wool of sheep and the hair of goats, and can protect himself from wind, the weather and the cold by using wood from the forest to build his house and to warm it.

Food, clothing and shelter are the fundamental requirements to which everyone born has a claim, and these can also be acquired by every one who has sound limbs. In the muscles of our arms we possess the fairy charm which can say: "Table be set!" for labor always finds its reward. Of course if people are foolish enough to leave the places where the muscles of their arms are in demand and paid for, if they leave the source of all earthly riches, agriculture and go where their arms have no value, because many others that are unemployed are waiting for employment, then distress, lack of food, of clothing and shelter must give him the occasion to consider and turn back, returning to a life in the country, which is continually becoming more deprived of its inhabitants.

Every work brings its reward. Work is necessary for our bodily and mental well-being. By co-operation it confirms us in the consciousness of a common humanity, for in social life we see in every fellow man an image of ourselves and this calls for mutual regard, charity, kindliness, mutual assistance. How different with the man who is not working. His thoughts turn to laying nets and setting traps in which to catch his unsuspecting fellow men.

Further, when the knowledge will have spread more and more that the essential work of man consists in allowing the sun to work for him, in order that food, raiment and wood may grow up from earth, water and air, then many foolish outbirths of idle brains will lose their soil and foundation.

There are, indeed, in these times some bad calculators who say: We will work less and get more money. These do not consider that the more money is in circulation, so much more money must be paid for the materials of food, if these remain the same in quantity, and this change will be of indefinite limits. The real remedy can only consist *in producing more food*. The more grain is raised, the less money will be required to buy it. Here we must apply our lever. What infatuation, when men attack one another in order to compel the supply of sufficient food. That can only be furnished by the earth. "Does a cornfield grow in my palm?" God has created us rich enough in supplying us with an understanding. If we use this, brother need not overreach brother, but we can in serene tranquility of soul win the little that we need day by day from our all-mother Earth.

STONE-MEAL AS A TOBACCO FERTILIZER

Of late years the general attention of tobacco growers centered in the query "What is the beat manure for obtaining a good tobacco?" For it stands to reason that, if for a number of years tobacco is grown on the same fields, in the course of time the soil must be rendered bare of the constituents entering into the remarkable quantity of ash which tobacco contains. There is no other product of the soil which gives as much ashes as does tobacco, for the best dried leaves will yield from 14% to 27%, while, for example, dried ash or beech leaves only yield 4,75%, and most other plants contain still less, dried pine needles only 1,25%. In the ash of most plants yielding 2% or more silex predominates ash and beech ashes containing over 33%, while the ash of barley and oat straw consists 50% of silex. It is, however, quite different with tobacco ash, which contains only one 2% part of silex, the rest being lime, magnesia, potash,soda, phosphoric and Sulphuric acid. There is no fixed rule in the proportion of these substances, but lime and potash always predominate in about the proportion of five to four parts.

German tobacco yields less ash than Virginia leaf, only about 14%, and consists of about five parts of lime, four of potash, one of magnesia, one-half of soda, two-thirds of phosphoric acid, four-fifths of sulphuric acid, four-fifths of silex and one part of muriatic acid.

The less of sulphuric and muriatic acid a tobacco contains the freer will it burn and the whiter its ash will be. The beat tobacco is raised with nothing but wood ashes for manure, and be it noted that the ashes of oak, beech, birch, pine and fir contain not a trace of muriatic acid and but 0,02% of sulphuric acid. We are forced, therefore, to the inevitable conclusion that the comparatively high percentage of sulphuric and muriatic acid which the ash of German tobacco yields and which makes its present quality so poor is owing to the persistent use of *stable manure,* and it is plainly of the highest importance to do without that altogether.

The question now arises what shall be used in its stead? Our answer, is that inasmuch as forest trees are grown on rocky soil which contains *potash, soda, lime* and *magnesia* in combination with *silica alumina* and *phosphoric acid* we must, instead of burning the expensive trees for the purpose of obtaining their ashes for tobacco manure, go back to the original substances out of which the trees were created, and these are suitable minerals found in the rocks. This is as plain a proposition as the egg of Columbus.

With regard to Virginia tobacco a study of the topographical features of the tobacco lands will be in order. The best soil for the purpose is found where the debris of the Alleghenies and their foot bills the "blue mountains" has been washed down into the plain. These mountains contain gneis, granite, syenite, serpentin and hornblend slate. Hornblend is silica combined with lime, magnesia and iron. In syenite lime and magnesia predominate over potash and soda; Virginia gneis abounds in lime, magnesia and potash; serpentin is a silicate of magnesia and iron. These lime and magnesia silicates are of far more importance for the production of a fine tobacco which will burn freely making a white and firm ash than the potash which is found in all primitive rocks, although potash is necessary for the production of elastic leaf cells so much appreciated in good tobacco. But it is a great mistake to lay undue stress on an overabundance of potash. Neither the Strassfurt potash salts nor powdered iron slag will produce good tobacco. For the potash contained in tobacco is not combined therein with sulphuric and muriatic acid, but enters into direct combination with cell material, and it is eliminated out of silicated potash and soda by the action of the carbonic acid of the air or of the soil. A healthy and fine quality of tobacco can therefore only be grown by the use of a liberal supply of a mineral mixture which yields in appropriate proportions silicate of potash and soda together with carbonate of lime and magnesia and a small proportion of phosphoric acid, such as was present originally in the virgin soil of the tobacco lands of Virginia.

In accordance with these principles suitable mixtures of the several kinds of rocks have been prepared in the form of very fine powder for the production of fine tobacco, and it is at present being used with great success in the Palatinate in Germany.

A PAPER CONTRIBUTED TO THE "DEUTSCHES ADELSBLATT", January 31st, 1892

In cereals, in the seeds of the leguminous plants, and of the oil-bearing plants, the mineral substances with which the cellular tissue and the vegetable albumen are combined constitute from 17 to 50 thousands. After the combustion of plant tissue these mineral constituents remain behind as ashes, and the greater part of the ashes in the seeds consist of *phosphoric acid* and *potash*, while soda, lime, magnesia, hydrochloric acid, sulphuric and silicic acid with manganese; iron and fluorine are comparatively less in quantity. Only in the oil-producing seeds (mustard, rapeseed, linseed, hemp-seed and poppy-seed) lime and magnesia make a considerable part of the ashes. The following numerical proportion will give a general view:

Winter wheat has on the average 16,8 thousandths of ashes, of which phosphoric acid forms 7,9 thousandths and 5,2 of potash.

Field beans yield 31 thousandths of ashes, of which phosphoric acid forms 16,2, potash 7, lime 18 and magnesia 5 thousandths.

Poppyseed gives 51,5 thousandths of ashes, of which 16,2 are phosphoric acid, potash 7, lime 18 and magnesia 5.

From the fact that phosphoric acid and potash have such a prominence in nutritive crops, it was easy to draw this conclusion: "*That potash and phosphoric acid are the most necessary fertilizers, and the more phosphoric acid the better.*" But this conclusion is erroneous and has caused us much injury since Justus(?) Liebig made this statement.

Liebig and his successors have overlooked the fact, that in the time of vegetation phosphoric acid is so uniformly distributed that it does not amount in the average to more than 10% of the mineral constituents. If during the process of ripening phosphoric acid strongly accumulates in the seeds, that it constitutes not merely 10, but 30 to 50% of the ashes, this is explained by the fact that the acid passes from the stems, stalks and leaves into the seeds, leaving the straw very poor in phosphoric acid, as may appear from these proportions:

(a.) *The straw of winter wheat has* in the average 46 thousandths of ashes, of which only 2,2, thus about 5%, consist of phosphoric acid. The rest consists of 6 potash, 0,6 soda, 2,7 lime, 1,1 magnesia, 1,1 sulphuric acid, 0.8 hydrochloric acid and 31 thousandths of silicic acid. The latter (silica) only amounts to 0,3 of one thousandth in the wheat grain thus in comparison with the straw only one thousandth.

(b.) *The straw of field bean* furnishes 45 thousandths of ashes, of which only 2,9 are phosphoric acid, thus 6,5% while in the ashes of the seeds it constitutes 36 per cent. The other substances contained in bean straw are 19,4 thousandths potash, 0,8 soda, 12 lime, 2,6 magnesia, 1,8 sulphuric acid, 2,0 hydrochloric acid and 3,2 silicic acid. On account of this small quantity of silica bean straw is *soft*, while wheat straw, rich in silica, is *hard*.

(c.) The straw of *poppy* gives about 48,5 thousandths of ashes, of which there are only 1,6 of phosphoric acid; i. e., in *poppy straw* phosphoric acid constitutes only 3,3% of the ashes, while in the seeds it amounts to 33%. So considerable, amounting to the tenfold, is the difference. The rest of the ashes of the straw of poppy consists of 18,4 potash, 0,6 soda, 14,7 lime, 3,1 magnesia, 2,5 sulphuric acid, 1,3 hydrochloric acid and 5,5 silicic acid.

The examples adduced are to a certain degree typical of cereals, leguminous plants and oil-yielding plants and they explain why leguminous and oily plants need more lime in the soil than cereals. On the whole, when we take the average of 70 or 80 analyses of field-crops, which also include the roots, stems and leaves, we come to the conclusion that phosphoric acid constitutes about 10% of the mineral constituents, while potash, soda, lime, magnesia, silica, sulphuric acid, chlorine and fluorine contribute the remaining 90%. Furthermore, potash and soda are present on the average in the same amount of weight as lime and magnesia. These four bases amount to about 80% of the whole quantity of the ashes, and it is found in practice that these bases may to a considerable degree act as substitutes for one another, without perceptibly varying the form and the organic constituents of these plants.

According to these facts a fertilizer which would satisfy the natural demand of supplying the minerals necessary for the construction of plants should contain to one part of phosphoric acid eight parts of potash, soda, lime and magnesia, if we are willing to leave out of our count phosphoric, hydrochloric and silicic acid.

Such a fertilizer, however, is found in every primitive rock. Primitive rocks do not, indeed, contain more than one 1% of phosphoric acid, but that is quite sufficient; it is, indeed, the measure wisely appointed by the Creator of all things, for the other constituents of granite, porphyry, etc., which serve for the nourishment of plants, consist of about 6% of potash and soda and 2% of lime and magnesia. The residue of the rock serves as a substance dispersed between the basic substances to keep them apart, and they are dissolved out of their combination with silicic acid only as they are applied to use. Thence we receive such wholesome cereals from mountainous countries; *e.g.*, from Hungary, encircled by the Carpathian Mountains, in contrast with the prevalence of diseases due to the decomposition of the blood of men and of animals in the exhausted plains which are supplied with stable manure.

If we wish to grasp quickly and completely the correctness and importance of mineral manure, we need only to consider the cases of Uruguay and Argentinia or of Egypt; or, to mention an example from our proximate vicinity, that of the principality of Birkenfeld.

In Uruguay and Argentinia the live stock is estimated at about 32 millions (beeves, sheep and horses). Of these there are now killed for export every year about 1,25 millions and the bones of these animals are carried by the shiploads to Hamburg, in order to be worked up into bone-black to be used in the sugar refineries. It is self-evident that the animals take the phosphate of lime for their bones and the nitrogen for their flesh and for the glue in their bones from the grass they eat. But the grass draws the necessary nitrogen from the air, for they use no fertilizers, and the phosphate of lime, which continually passes from the country in the form of bones is received by the grass from the inexhaustible calcareous porphyritic mud which is carried down through millions of gorges from the Cordilleras by the mountain streams and which flows as a primitive manure into the eastern plains. In Egypt this is effected by the Nile mud, which the mountain streams bring down and which is conveyed by the Nile in fructifying abundance to the Delta, which thereby becomes the granary of Egypt.

But we need not go so far even. The little principality of Birkenfeld demonstrates the fertility of the primary rocks which the mountains of the Hundsruecken supplies in the form of argillaceous slate. It is a little Argentinia. The trade in cattle plays an important part in Birkenfeld. Besides this oil factories, linen factories and beer breweries prove that cereals and oil plants, rich in phosphorus, and among them flax, rich in potash, find there good nutritive supplies. The forests consist mainly of deciduous trees and harbor much game. Trees need phosphoric acid for their roots, trunks and bark, and the game needs phosphate of lime for the bones. The ashes of oak wood and beech wood contain 6% of phosphoric acid, and that of the horse-chestnut contains 7%. So richly does the argillaceous slate furnish the nutritions elements for the growth of plants and especially the right quantity of phosphoric acid.

In contrast with these natural fertilizers what has our prudent and learned fertilizing with phosphoric acid effected? It has brought it about that we don't know how to save ourselves from the phylloxera, the nematodes, hay-worm, spring-worm and sour-worm nor from the fungi-causing rust and blight. The more phosphoric acid the more parasites, because fungi and parasites need the phosphatic protoplasm which accumulates in seeds and fruits as an essential condition of their existence. If we wish to limit these plagues to a sufferable degree we must supply our fields that have been deluged with phosphoric acid with natural plant-food, with pulverized rocks, with lime and gypsum.

Of many communications received which confirm the above, we would like to cite a few which are especially instructive, as it shows that these evils have become so great as to urgently demand relief. The representative of a great vineyard estate on the upper Rhine writes as follows:

"For years I have seen clearly that we make a great mistake in our cultivation of fields, gardens and vineyards, but only on reading your books have I seen that all our methods of fertilizing hitherto have been one-sided, and that, therefore, they are ineffectual. Stable manure on some soils and for some crops may be sufficient, but it is not a universal fertilizer. We see this plainly here in the Rheingau, in the young vines, which are manured every two or three years with cow dung, and, indeed, great quantities of it. A gladsome, luxuriant growth and a rich yield of grapes are *not* produced, though we furnish the grape-vines with the potash, phosphoric-acid and nitrogen in so great quantities that the shoots, the grapes and the leaves ought to display the most luxuriance; but instead of this everything in the vineyards here looks sickly and poor. I should, therefore, be very glad and grateful to you if you would give us your views about this. It would be a great benefit, not only to ourselves, but to the whole of the Rheingau, and wherever grape-vines are cultivated, to be delivered from the miseries of the spring-worm, hay-worm and sour-worm, the phylloxera and the *Peronospora viticola*, and if this can be done by your method all cultivators of the grape-vine will exclaim: God be praised!"

I answered that the usual manure does not lack any necessary ingredient, but there is in it *too much of some things; i.e.*, of nitrogen and phosphoric acid. Men must return to the original material, restore to the soil its natural original qualities by bringing to the fields soil that has not been exhausted, which may be done in the form of powdered primitive rocks mingled with sulphates and carbonate of lime and magnesia. The correctness of such belief is attested by the following correspondence with a landscape-gardener and nursery-man from the Rheinprovinz:

"We would like to ask you for some information as to what we had best use for manuring our nurseries. We have clayey, deep, light soil, formerly a forest. We cultivate roses, fruit trees and forest trees, also evergreen plants, firs and various kinds of cypresses. It is quite peculiar that quinces and other fruits (Formobst) in the second year after grafting absolutely refuse to grow any more despite of the use of stable manure, iron slag or of Chili-nitre."

I answered that deep, clayey forest soil while retaining its clay and silica has been deprived of its basic constituents (potash, soda, lime and magnesia) which in the process of time have passed over into the wood of the roots and of the trunks, and that the only thing promising relief is fresh rock meal. For are not the Balkan countries the home of the roses, and do not the Haemus Mountains consist of porphyry, granite and gneiss, but not of stable manure and clay? Do not cypresses grow in the regions of the Apennines, which furnish the nourishing material from their granite and gneiss? And do not firs grow on mountains of granite and porphyry? Finally fruit? The Bohemian Mountains furnish it in abundance, and indeed free from worms. This latter fact, that the use of stone-meal causes worms to cease, was lately confirmed by Mr. Fischer, M. D., of Westend, near Charlottenburg, who introduced stone-meal manure two years ago in his garden, situated on a sandy soil. He reported about it in the January number of the "Deutsche Pomologen-Verein".

From a third letter I quote as follows:

"Manor L. - I am glad to see a chemist who has the courage to openly oppose the swindle of the artificial manures. Within a series of ten years I bought at least $17.000,00 worth of artificial fertilizers, of which sum over $6.000,00 were paid for Chili-nitre. I harvested more every year; but what? Nothing but straw, lodged grain and cereals of low grade. For the last two years I have bought, in addition, animal manure and lime, and I find that at a slight expose everything is being changed and that the field will again bring in what I lost in former years. When the Thomas phosphate was introduced, as it was cheap, I used at once 2.000 cwt. (100.000 kg). With 7 cwt. (350 kg) por acre an effect was indeed seen, but what was it that acted? Surely only the lime. What you have affirmed I have long felt. That many of us agriculturists are faring so badly is for the most part owing to this nuisance of our artificial, expensive and useless fertilizers."

A fourth letter with an excerpt, of which I will conclude, contains the following:

"Twenty years ago, while in office in Alsatia, I endeavored to make myself acquainted and familiar with all manner of subjects. I was lead to the idea of mineral fertilizers or manures, when I heard and saw that in the intersecting valleys of the Vosges Mountains the winter torrents covered the lowlands with granitic debris, which after a few years became very fruitful soil; but I had no opportunity or occasion to follow out this idea any further, which is now, however, the case." (G. L., Privy Councillor of war a. D.).

Every such letter contains new confirmatory facts; I have quite a collection of such correspondence, but will not weary you by quoting more.

JULIUS HENSEL
Hermsdorf unterm Kynast

STONE-MEAL MANURE
"Pioneer", July 22, 1892

> *"Bread from stones: and thus forsooth*
> *- The Bible words maintain their truth"*

I have before this taken occasion, in the "Deutsche Adelsblatt", to show that calling the stone-dust "manure" is really not correct, as it is superior to the so-called manures in this *that it restores the natural conditions for the growth of crops*, while manures only present an artificial help and thus a makeshift. The whole state of the case is as follows:

In the beginning plants grew without any artificial addition from the soil formed of disintegrated material from the mountains. The carbonic acid of the air combined with the basic constituents, potash, soda, lime, magnesia, iron and manganese, which were combined in the disintegrated rock-material with silicic acid, alumina, sulphur, phosphorous, chlorine and fluorine, and with the co-operation of moisture by the operation of the heat and light of the sun it produced vegetable cell-tissue. The gaseous substances, *carbonic acid* (carbon dioxide), watery vapor and the nitrogen of the air acquire the firm forms of vegetable cellular tissue and vegetable albumen solely through the basic foundation of potash, soda, lime and magnesia, without which no root, stalk, leaf or fruit is found; for whether we burn the leaves of maples or of beech trees, the roots of burdocks or of willows, grains of rye or wood, straw or linen, pears, cherries or rape seed, there always remains a residuum of ashes which, in various proportions, consists of potash, soda, lime, magnesia, iron, manganese, phosphoric acid, sulphuric acid, fluorine and silica. With respect to nitrogen this with watery vapor forms in the presence of iron, which is present in all soils, becomes ammonia according to the formula $N_2 H_6 O_3 Fe_2 = N_2 H_6 Fe_2 O_3$ (all iron-rust that is formed in the nightly dew out of metallic iron $Fe_2 O_3$, contains ammonia, as Eilard Mitscherlich has proved). The solidification of the cellular tissue arising from carbonic acid and water will be best understood by comparing it with the process of the formation of hard soap by the combination of oil with soda, potash, lime or any other basic substance, as, *e.g.*, oxide of lead, quicksilver or iron. Ammonia also forms soap with oxidized oil, oleic acid. We can hardly find any better comparison by which to explain the solidification of the atmospheric vapors (carbonic acid, water, nitrogen and oxygen) in combination with earthy substances or in substitution for the latter with ammonia into vegetable substance than on the one side this process of saponification and on the other hand the oil substance which is the basis of soap. The production of oil substance consists in this that combustible substances (hydro-carbons) are generated from burned-up substances (carbonic acid and water) and this characterizes in the main the nature of the universal vegetation of plants. A burning stearine candle is transformed into carbonic acid gas and watery vapor, but these aeriform products, in combination with earths, are again transmuted into combustible wood, sugar, starch and oil by the operation of the sun. Wherever new earth comes into activity, as at the foot of mountains, there is found a vigorous growth of plants, especially when a sufficiency of carbonic acid clings to the rock as in the Jura regions. The road from Basel to Biel is very instructive in this respect. On the contrary, it is seen that in densely populated regions as, *e.g.*, in China and Japan, after a cultivation of many thousands of years, the earth, exhausted of the material that forms cells, is of itself unwilling to produce as many nutritive plants as men and animals need for their sustenance; but as had been perceived that the nourishment which has been consumed, in so far as it has not been used in the new formation of lymphatic fluid and blood, being therefore superfluous, leaves the body through the digestive canal although chemically disintegrated and putrefied, nevertheless produces new vegetation when this material is brought on the fields and is mixed with earth; in China they collect with great care not only whatever has passed through the intestinal canal, but also the product of the bodily substance which is consumed by respiration, which is eliminated as the secretion of the kidneys and which also gives an impulse to new growth.

One or the other must take place. Either unexhausted new soil or the restoration of the nutrition consumed to the soil of the fields. Where the latter has not been done, as by the first European settlers in America, the crops decreased and the settlers moved from the east further to the west, in order to gain enough cereals from the as yet unexhausted soil for export to Europe. Now they have also come to see in America that they cannot continue this, as there are no more domains without owners into which they can emigrate without let or hindrance.

But how is it with us in Germany in this respect? After the soil would not yield any more, despite of deep plowing, the circle instituted in China was also put into practice. They had to see that the solid and liquid manure of the domestic animals brought on the fields produced a new growth, and the dung heaps began to be valued. By the aid of this dung the fields were kept fertile, although this was a mere makeshift. This makeshift has become a familiar one for several centuries, so that even in the times of our great-grandfathers the saying was in vogue: "When there is no manure nothing will grow." So eventually what was a mere makeshift has become the regular rule. As a consequence of this traditional view the conclusion followed: In order to get a large quantity of manure we must keep as many cattle as practicable. In this it was overlooked that the cattle would require again as much acreage for their nutrition, and the ground thus used could not be used to raise grain, so that in such an economy it was necessary to work the fields for the sake of the cattle not for the sake of the men. But finally the thoughtful and book-keeping farmers had to come to the conclusion that the raising of cattle only pays in mountainous districts, or in districts like the marshes of Holstein, which are kept fruitful by the continual washing down of Geest-rocks.

I can only summarize here. As above said, the dung heap had been recognized as the augmenter of fertility, and dung heaps were considered as the natural condition *"sine qua non"* for the growth of crops, although this was by no means founded on the natural Order, but was only a makeshift. When once the rule was established that the artificial was normal we need not be surprised that when the stable manure would no more suffice some people recommended *artificial manure*. As these people gave themselves great airs of learning the well-educated, large land-owners fell into their net, even more than the simple peasants, and therewith the general retrocession of agricultural produce in the level regions was for some time at least fixed and sealed.

It may easily be seen that oxen and cows, no matter how high their cost, charged no salary for producing their manure. It was otherwise with the chemists and the dealers in artificial manure. These not only demanded to be nourished themselves, but also desired from the gain produced by their business to educate their children, to build their magazines, to pay their agents and to increase their capital. This business like all those which supply necessaries proved so remunerative that one of the greatest houses dealing in artificial manures in a short time had made millions, which were paid them by the farmers without receiving an equivalent; for in spite of the most energetic application of artificial manures the crops steadily decreased. How could it be otherwise? Plants need potash, soda, lime, magnesia, iron, manganese, sulphur, phosphorus and fluorine, and in the artificial fertilizers they only received expensive potash, phosphoric acid and nitrogen for their nourishment.

The consequence of this showed itself first of all in frequent bankruptcies of agriculturists. But besides this, nitrogenous fertilizers in the form of Chili-nitre have caused a predominance of cattle diseases. That hares and deer have been found dead in numbers in places which had been fertilized with Chili-nitre I have read in at least twenty newspapers, and it has also been reported to me by eye-witnesses. As in the open air so also in the stables. No normal animal bodily substance can be formed from fodder manured with nitrogen, especially no wholesome milk equal to that from cows feeding on mountain herbs.

It is not to be computed how great an injury to health with men and animals has been caused by stable manure. Milk produced from ammoniacal plants paved the way by which the destructive spirit diphtheria has swooped down after measles, scarlatina, scrofula, pneumonia had become the familiar companions of the Germans, who before were strong as bears. Artificial manure at last put the crown on this work of destruction.

How could this happen? Very simply. Liebig was the first agricultural chemist. He found that the ashes which remained from grain mainly consisted of phosphate of potash. From this he concluded that phosphate of potash must be restored to the soil, and that was very one-sided. Liebig had forgotten to take the straw into account, in which only small quantities of phosphoric acid are found, because this substance during the process of maturing passes from the stalk into the grain. If he had not only calculated the seed but also the roots and the stalks, he would have found what we know at this day, that in the whole plants there is as much lime and magnesia as potash and soda, and that phosphoric acid forms only 10% of the sum of these basic constituents. Unfortunately Liebig also was of the opinion that potash and phosphoric acid has to be restored to the soil as such, while any one might have concluded that instead of the exhausted soil we must supply earthy material from which nothing has been grown. Such untouched earthy material of primitive strength we get by pulverizing rocks in which potash, soda, lime, magnesia, manganese, and iron are combined with silica, alumina, phosphoric acid, fluorine and sulphur. Among these substances fluorine, which is found in all mica-minerals, has been neglected by Liebig and by all his followers, and has never been contained in any artificial manure. But as we know from later investigations that fluorine is regularly found even in the white and yellow of bird's eggs, we must acknowledge it is something essential to the organism. Chickens get their fluorine and the other earthy constituents when they have a chance to pick up little slivers of granite. Where this is denied them, as in a wooden hen house, they succumb to chicken cholera and chicken diphtheria.

We men are not as well off as the birds of the heavens. We must eat the soup prepared for us by the dealers in artificial manures. Since these sell no fluorine our cereals suffer a lack in fluorine, and as no normal bony substance can be formed without fluorine in the same degree as the number of dealers in fertilizers increased the army of dentists and the erection of orthopedic institutes increased; but the latter were unable to remove the curvature of the spine in our children. The enamel of the teeth needs fluorine, the albumen and the yolk of the eggs require fluorine, the bones of the spine require fluorine, the pupil of the eye also needs fluorine. It is not by accident that Homeopathy cures numerous affections of the eye with fluoride of calcium.

How rich, how strong and how healthy will we Germans be when we make our mountains tributary to yield new soil from which new wholesome cereals may be formed. We need then no more send our savings to Russia, to Hungary, to America, but will make our way through life by our strong elbows and with German courage, and shall keep off our adversaries.

The goal aimed at, of satisfying the hungry, and of preventing numerous maladies by restoring the natural condition for wholesome plant growth, seems to me one of the highest and the most noble. Even 6 cwt. (300 kg) of prepared stone-dust to the Prussian morgen (= 0,25 hectares, or about 10 cwt. to the acre) will give sufficient nourishment for a satisfactory crop, if this amount is supplied every year. If more is used, the yield may be so much the more increased.

I conclude these remarks, which were introduced with a motto that adorned the exhibit in Leipzig of the produce yielded by stone-dust, by reproducing also the second rhyme which had been introduced there, and which, like the motto, has a conscientious adherent of mineral manure for its author:

> *"Art we love, but never can endure*
> *To see the artificial in manure."*

JULIUS HENSEL
Hermsdorf unterm Kynast

CONTRIBUTIONS FROM OTHER SOURCES

STONE-MEAL
By Herm. Fischer, M. D., Westend, Charlottenburg
From No. 1 of "Pomologische Monatshefte", 1892, Edited by Friedrich Lucas, Director of Pomological Institute in Reutlingen

Not only those who like to eat fruit and vegetables, but much more those who raise fruits and vegetables rejoice in the abundant and savory produce of our gardens. To maintain this produce and, if possible, to increase it is the endeavor of rational horticulture. This end is striven for through careful cultivation, and more especially by abundant manuring, especially with nitrogenous compounds. I say this end is striven for, but it is not always reached. The long-continued labors of a well-known investigator, Julius Hensel, have opened new prospects for agriculture, fruit raising and horticulture; they show, in fact, how we can "turn stones into bread." Hensel's book, "Das Leben", has lately appeared in a second edition. Every thinking reader will find a high enjoyment in the study of this book. For our present consideration I recommend especially Chapter XXX., p. 476, "Agriculture and Forestry". Lately a little work, by the same author, has appeared on "Mineral Manure the Natural Way for Solving the Social Question," published by the author at Hermsdorf unterm Kynast, Silesia. The first part of the pamphlet is devoted to the defensive, for like all pioneers our author meets with violent opposition from the orthodox teachers of agriculture, whose cues and periwigs have come into a great state of agitation.

After his defense the author passes to his theme proper. Earth, air, water and sunlight must co-operate to produce a fruitful growth. We entrust our seeds to the earth. What is the "earth?" The earth or soil is disintegrated primitive rock (gneiss, granite, porphyry). The soil of our fields is continually being increased by the disintegration of primitive rocks, and from this there grow up grasses, herbs, shrubs and trees; without mineral constituents no plant can grow. Now, when in level plains the upper layer of the soil through long cultivation has become exhausted of certain necessary mineral constituents new rocky material must be provided, from which nothing has as yet been grown, which, therefore, still contains all its strength; this is not only the most natural, but also the simplest and at the same time the cheapest way to increase and maintain the yield of our fields. This is not mere theory, thought out in the study; but experience and success have demonstrated it. With Hensel there is no more need for experiments, but merely of demonstration. A firm has produced a variety of fertilizers, according to his directions, out of pulverized rocks, such as are most suitable for the various plants. I will here only mention fertilizers for vineyards, meadows and potato fields. Hundreds of advocates affirm the favorable results of these fertilizers. The rest should be read in the pamphlet itself.

Since the spring of 1890 I have used stone-meal manure in my garden, situated on our well-known sandy soil, and am extraordinarily well pleased with the result. I have, *e.g.,* picked from a row of raspberry bushes about 23 yards long fifty quarts of the most delicious fruit, some of over one inch in length and 75% of an inch in diameter. The shoots of this year, which will bear next year, are as thick as a finger, some as thick as a thumb, and up to eight feet high. The young fruit trees planted about three years ago are bearing very well, and what is well to notice they are set abundantly with buds for blossoming next year. What is especially surprising is that I have found no worms at all. Neither in my raspberries nor my early pears and apples. The winter apples also have so far not shown a single worm-eaten fruit.

My vegetables I sowed in furrows, covering first with mineral manure and leveling the furrow with earth. The plants I took out to transplant had a mass of roots such as I have never seen even in a manure bed. They, therefore, were easily transplanted; none withered. I will not mention my asparagus because the variety used (Horburger Riesenspargel) of itself brings great shoots. I have cut asparagus weighting six to nine ounces; they were a foot long and their circumference at the middle of their length was 4,5 inches; the taste of this asparagus is excellent. I would specially point to the quality, the most delicious savor of fruits, etc., grown with this manure in contradistinction to those grown with stable manure; this is also shown in the pamphlet mentioned above. With all these advantages mineral manure is even cheaper than all other artificial manures. "We need no artificial manure if we supply that which we annually draw from the soil in form of fruits, etc., by means of fresh, unexhausted pulverized granite, gneiss or porphyry as the genuine strengthening and primitive fertilizer, mixed with gypsum and lime."

How the fungus of the grape-vine, the Odium Tuckeri, is to be removed and how even the phylloxera can be extirpated and, according to Hensel's statements, has been extirpated, may be seen in "Das Leben," p. 478.

The fallacy of the supposition hitherto held that all cultivated plants must have especially nitrogenous food in order that they may prosper becomes more and more apparent. By experiments it has been indubitably proved, and Hensel always asserted, that plants, and especially the leafy, leguminous fodder plants (clover, vetches, etc.), can take up and elaborate nitrogen through their leaves out of the air just as the carbonic acid taken up from the air is worked up into hydrocarbons under the operation of light. All we need, therefore, is to furnish the soil with the necessary mineral constituents. Mineral manure is the most profitable, most lasting and, what is not to be overlooked, an entirely odorless fertilizer.

If i shall have succeeded in calling the attention of the reader to the glorious effects of this manure the object of these lines is attained. When the use of this manure is then followed by surprising results the beautiful fruits will, in the most literal sense, be my reward.

STONE FERTILIZING
By Dr. Emil Schlegel, Pract. Physician in Tübingen
From the "Wegweiser zur Gesundheit", Sept. 15, 1891

This is a subject that does not immediately concern the "Wegweiser zur Gesundheit", but which nevertheless on account of its far extended importance may have the greatest effect on the well-being and wealth of our people. The chemist, Julius Hensel, of whom we have several times before this spoken in earlier numbers of the "Wegweiser", and who is known to its readers by his genial book "Das Leben", has lately published another work which deserves particular mention. He therein sets forth that the loss of soil in mineral substances (lime, magnesia, etc.) is not supplied by animal offal, though this produces a strong forcing of the plants, which makes the leaves and the products weakly and injurious, as this is said to have developed in the irrigated fields at Berlin, where the bones and muscles of the animals fed on their produce are suffering and also the milk is not satisfactory for sucklings. In a still higher degree these injurious forcing substances are found in artificial manures and especially in Chili-nitre, causing a rapid, surprisingly luxurious growth; but when the fruit or the seeds develop there is a manifest falling off. Now, since every year millions of dollars are transferred from the pockets of the farmers into those of manufacturers of artificial manures, and of speculators and stock holders, this amounts to an impoverishment of the soil by parasites.

The true cure of an exhausted soil consists, according to Hensel, in supplying it with comminuted rocks, especially granite, gneiss, porphyry and lime. Thereby the plants receive again what they naturally demand. The "Wegweiser" would here remark that the best proof of these views given on a great scale is thousands of years old; i. e., the fertility of Egypt. The mud of the Nile consists almost exclusively of finely comminuted rocks, with very, very few organic nitrogenous constituents. But the flooded districts owe their unexampled fertility to just this precipitated stone dust. Hensel writes at the end of his book:

"Almost every field contains stones which have only been acted upon in part by the dissolving moisture of the soil, and which therefore shows a more or less rounded form. These stones, as they injure the spade or plow, are usually removed to the sides of the fields and there heaped up, and are then sold at a cheap rate for use on the highways. The farmer who acts thus sells his birth-right, so to say, for less than a pottage of lentils, for he removes the source of fertility from bis fields. If such stones are heated in the stove or on the hearth for half an hour and then thrown into water they become so friable that they may be broken into small pieces by the hands and may easily be pulverized with a hammer." It is to be wished that these developments of Hensel should find a wide diffusion.

LETTER TO MR. SCHMITT
Oranienburg, Aug. 17, 1893

Highly Honored Sir:

I have just safely returned from my long tour for the stone-dust, having been away five weeks, and I herewith give you a brief report, so that you may also enjoy the victory which stone-dust has gained wherever it has been really put into a practical test.

I have already written to you of the eminent, happy effects of stone-dust on the estates of Count Chamare. I have been able to see its good effects also in Upper Silesia, and have established there two more stations for the future, where normal trials will be made. I saw exceedingly significant results from stone-dust on the field of Chief Bailiff Donner at Culmsee, in West Preussen; i. e., excellent wheat, sowed after barley and oats, with only 5 cwt. (250 kg) of stone-dust to the acre; also splendid rye in fourth succession on 5 cwt. of stone-dust, and sugar beets following sugar beets on merely 6,5 cwt. to the acre, which promise a very good yield. Here it was found that, the fields needed above all a good supply of lime, and this lime was the best support to the happy effects of the stone-dust. On this account the cultivation of the field with stone meal demanded a simultaneous application of lime of 16 to 30 cwt. (800 - 1.500 kg) per acre.

So great a quantity will not be used in one year. For the stone-meal made according to Hensel's directions contains as much lime and magnesia as the average crops call for.

The cultivation of sugar beets can be doubled by stone-meal. This accomplishment would surely be a great result from stone-meal. Also in West Preussen I have established an experimental station for the proper use of stone-meal on a large estate near Braunsberg, belonging to a Herr von Bestroff. This gentleman called on me for this purpose also before this in Oranienburg.

I hope that this, my first tour in behalf of stone-meal, has not been in vain, and I intend, God willing, to repeat these tours annually, so as to benefit our great and important cause with all my strength. I am quite confident that stone-dust combined in the proper way with lime will by its practical success carry off the victory.

I shall do my best to carry out the stone-meal experiments on the estates of Count Chamare in the most conscientious manner, and hope that God's blessing may rest on this my labor, which I perform gladly for my country.

OTTO SCHOENFELD,
Director of the Agricultural and Forestry School

TO THE POMOLOGICAL SOCIETY "HEIMGARTEN IN BUELACH"
Letter by Mr. K. Utermohlen, Teacher in Leinde

By means of the stone-meal manure of Hensel we shall soon surpass all similar undertakings (co-operative Pomological Association). If the tree has a sufficiency of this primitive substance under its roots it is not only fruitful, but no more sensitive as to frost and diseases. Nor will it be infected as much by insects, as it will be healthy, having a pure sap. With the usual treatment with manure rich in nitrogen the trees are satiated to repletion, and then it is with them as with men. Their fibres are relaxed, their sap is checked, diseases developed, lice and other vermin infest them, and then we have to sprinkle them with mixtures, cut out wounds, put on wax and pitch, etc. By well preparing the soil with this mineral manure we prevent all these troubles from the start. The trees become strong and hardened. It is just as when parents bring up healthy children with solid food. They then have none of these troubles and cares encountered by parents who treat their children perversely.

For the last two years I have been making various experiments with stone-meal manure, and indeed with the different kinds. From my experience with it I have come to the firm conviction that we need no other manure at all but this. I wish I could speak with angels' tongues to make clear to you its great importance for our cause. It would carry me too far to speak of all the various experiments. A radical reform in this direction will have to be made. If we give the trees when they are first planted some of this manure between their roots, with good irrigation, they will be twice as strong and vigorous as without it. We do not need any stable manure to loosen the ground, that is best effected by diligent hoeing and digging. Where this should prove insufficient we call in peat moss to our aid, and this can be gotten cheap here. That is what I did here with my heavy garden soil, and then with the help of stone-meal I have raised the finest vegetables, though the garden has seen no stable manure for 8 years. And then how pleasant and cleanly is this mineral manure when compared with the smell of solid and liquid stable manure. Then we should consider its great cheapness. Much can be done with 1 cwt. (50 kg). If we had always to use stable manure we would have to give out great sums every year, and even then we could not get a sufficient quantity. But there must be manure, for "from nothing nothing comes", as the saying is. In this trouble the mineral manure is our best help. We cannot in this matter give any consideration to the authorities in horticulture, as they are in error with respect to the nutrition of plants. I refer especially to their silly theories about nitrogen. Who brings to the strong oaks of one hundred years growing on rocky soil, or to the other lovely children of mother earth out in free nature, liquid or solid manure or sewerage? They grow and nourish and revel in their healthy growth just because they are spared all these. So it will be with our fruit trees when we shall nourish them in a natural manner. It is not a mere secondary question but a most fundamental one which is here involved: The question is whether we shall in the treatment of our little trees follow the perverted and worn out routine of the wisdom of the professors of our state with their theories of albumen or whether we will follow the path of nature. We have chosen for ourselves and our mode of living the latter course; it is then surely proper to do the same with respect to our plantations.

If I only had a photographic apparatus! I would like to send you pictures of some of our standard trees and some of our half-standards, so that you could convince yourself with your own eyes of the excellent effects of this wonderful fertilizer. This in especially the case with a four-year-old half-standard to which I have specially applied this manure. Such a multitude of the finest russets! It would hardly be thought possible in a little tree of four years. And then you should see how this little fellow has increased in thickness! His coat has almost become too narrow for him. The apples hang twice as thick as in other years, and their flavor can hardly be recognized; their aroma is really refreshing. The same I have perceived this year in our cherries and raspberries. When I come to see you I shall bring a whole selection of apples for trial. I well manured a bed of several square yards of ground and planted it with cucumbers.

After gathering this summer a whole basket full I thought I had a remarkably good crop; but now the bed is just as full again, although I have picked some from time to time. The same is the case with the beans and onions which I have noticed particularly, as we can only plant flat-rooted vegetables between the trees.

We cannot sufficiently express our satisfaction that we have in this manner not only found a substitute for, but something far better than stable manure.

THE STONE-MEAL OF DR. HENSEL BEFORE THE COMMITTEE ON FERTILIZERS OF THE GERMAN AGRICULTURAL SOCIETY (DLG)

From Dr. F. Schaper, Nauen, in "Osthavellaendisches Kreisblatt"

"Most of the members evidently knew nothing about the mineral manure save through the abuse of the well-known Professor Wagner, in Darmstadt. It is a sad state of affairs, but it is true, that these institutions, founded for the use of agriculture, cannot act freely, but have to regard quite different groups of interest; i. e., those of the manufacturers of manure. That their interests and those of the farmers are directly opposed to each other is manifest from this that farmers desire cheap fertilizers but the manufacturers of manure desire to keep them as high as possible in order that they may make the more money. Now the agricultural trial stations receive part of their support from the manufacturers of manures, as they are paid for their control-analyses, experiments, etc. In order that they may not lose these contributions these institutions must avoid whatever runs counter to the interests of their employers. It is often even stipulated in the contracts between the manufacturers of fertilizers and the agricultural trial stations that they should obligate themselves to protect these factories of artificial manures from unfair competition.

But who is to decide who and what belongs to "unfair competition"? The manufacturer will be apt to consider every one as an "unfair competitor" who threatens to diminish his profits, and he will therefore insist, and a certain plausibility cannot be denied to their demands, that the agricultural stations according to their contract should in every case work for them. This enables us to explain the silence or the open hostility of the agricultural trial stations as to stone-meal manure.

No intelligent man will on this account consider this hostility of importance or take too serious view of it. This opposition should even be of use to the cause, since no truth valuable in itself can be injured by the exercise of a criticism ever so sharp, if this is done in a scientific manner. But such an objective criticism has not been exercised on Hensel's theory, but certain directors of trial stations, instead of combatting it in a scientific manner, have descended to gross abuse and have, therefore, been judicially punished.

Mr. Shulz-Lupitz, the chairman of the Committee on Fertilizers, objects to Mr. Hensel, in the session of 14 February of this year (1893), that he is conducting his cause against acknowledged men of science in a rough manner, and that this could not be rebuked sufficiently - a peculiar objection as coming from a man who, so far as the direction of the proceeding and the form of the resolution offered by him and finally accepted go to show, him only a slight regard for the whitewashed politeness of Europe. He has, we are sorry to see, forgotten that Mr. Hensel was not the attacking party, but quite a different set of people, the close friends of Mr. Shulz-Lupitz, and the aim of the proceedings was evidently to get them out of the scrape into which their own precipitation had brought them. The well-known professor, Dr. Wagner, in Darmstadt, director of the agricultural trial station there, in his edict in the year 1889 had called the mineral manure a gross swindle and denied to it any value. This edict had been published by "Zimmer's factory" in Mannheim in innumerable pamphlets and in journals as a supplement. Thus it came that in far extended agricultural circles which only heard of mineral manure through journals of Wagnerian tendency Mr. Hensel was accounted a charlatan? When a man like Mr. Hensel who thinks he has discovered something useful for agriculture is thus shamefully reviled, and in the end deals with his assailants in a somewhat doughty fashion, who will account him reprehensible? Now Mr. Schulz-Lupitz in the proceedings continues this kind of polemics against Mr. Hensel.

The resolution passed declares in its first part: "Hensel's stone-meal is from the standpoint of practical and scientific knowledge to be designated a worthless fertilizing agent". Just the contrary is the truth. From the standpoint of practical experience the stone-meal has shown itself a valuable fertilizer; surely enough, the men who had some practical experience in the manure were not acknowledged by these gentlemen of the manure division! But they were presented by some learned men of this assembly (Latin economists Uncle Braessig would probably call them) conscious of their infallible book learning, as men who could easily be cheated, and who now also cheat others, thus as cheating and cheated.

These learned gentlemen seem to forget that in practical life a grain of common sense outweights a hundred weight of book learning, as the shepherd of the Abbot of St. Gall said long ago.

In the second part of its resolution the manure division rebukes the "impertinent bearing" of the "so-called chemist" Hensel with indignation and "expresses to Mr. Professor Wagner, in Darmstadt, the thanks of the practical agriculturists for his appropriate designation of the stone-meal of Hensel. This latter gentleman had called it an above mentioned, "a gross swindle." The manure division has cautiously avoided using this expression. For this expression has caused the punishment of two editors who had copied the Wagnerian production and its author. Professor Wagner has escaped a probable judicial condemnation only by the fact that the complaint owing to an "oversight" fell under the statute of limitation.

We who are convinced of the value of Hensel's method of improving the soil look trustingly into the future, with the conviction that truth has always made its way if it only found courageous and intelligent champions.

I would, therefore, request all who have had any practical experience with the stone-meal to publish their experiences for the good of the cause and of their fellowmen, and not to leave the field to the sole occupancy of the opponents. The word of the *single* man easily dies away, the *multitude* only makes the full chorus, especially in our democratic times, and this chorus alone can hush the short-sighted insolence and the self-interests which oppose the new discovery.

ABOUT STONE-MEAL MANURE
From the "Land- und Hauswirthschaftliche-Rundschau"
No. 11, 1893

A short time ago we published an article on the experiments with the new stone-meal fertilizer; we also gave space to an objective presentation as to the causes which make stone-meal suitable for a manure. The new fertilizer and its discoverer have suffered severe infestations. It may, therefore, interest our readers to see a report from our neighborhood as to some trials made of it. We have received the following:

Sometime ago a burgomaster of the neighborhood called our attention to the splendid stand of grain manured with stone-meal on the "Stenheimer Hof", on the estates of the Grand Duke of Luxemburg. A company of gentlemen who take an earnest interest in this matter (chemist, Dr. Ebel, teacher Eisenkopf, and the land proprietor, Loeillot de Mars, from Wiesbaden; Director Spiethoff, editor of the Pionier, from Berlin; Mr. Forke, of Eltville, and Dr. Dietrich and Dr. Broekhues, from Oberwallauf) in a excursion verified these statements beyond all expectations. In spite of the great drouth the rye on 18,5 acres of ground had stout stalks and long thick ears, and the tenant, Mr. Heil, told us that little more than 5 cwt. (250 kg) to the acre, altogether 100 cwt., had been used. Just as luxuriant with dark green stalks and leaves stood the oats, 1,5 acres, right by the highway. This piece of ground had not had any stable manure for many years, and had only received 20 cwt. of stone-meal with an addition of 6 cwt. of iron slag. The comparison with neighboring fields which had been well cultivated but differently manured, was very much in favor of the manuring with stone-meal. Just as striking as was the success of Mr. Forke on his rye, oats and clover, it was on his fruit trees and grape vines. We would only mention that a clover field of which one-half had been manured with stable manure and the other half with stone-meal showed a dense growth of clover on the latter half, while the former half showed many weeds but hardly any clover. A cherry trow and a tree with Gravenstine apples, which for many years had yielded no fruit worth speaking of, this year, after having been well supplied with stone-meal, are covered over and over with fruit.

A neighboring farmer told him, on seeing his fine oats: "Here we can see clearly how your manure acts; it could not stand better if you had put on 60 cartloads of stable manure per acre, which would have cost $125 to $150 per acre.

The condition of the grape-vines after repeated manuring with stone-meal was on comparison with other grape-vines found to be excellent, but we shall return to particulars, as with the rye and oats, at the time of harvest. We invite the farmers of the neighborhood to make their comparisons and to convince themselves of the solid results of manuring with stone-meal. This possesses the quality of *vigorously nourishing the plants and making them strong to resist frosts and drouth*. The above-mentioned gentlemen will bear record as to whether Hensel is really the "false prophet" that he has been represented to be.

To Director Spiethoff this investigating committee, in which he took part, was the more wished for, as the "Pionier" had first called attention to the scientist Hensel, and had also been the first to communicate last year the astonishing results in the Agricultural School of Oranienburg.

WHAT HELP CAN BE GIVEN TO THE HARD-PRESSED FARMERS
From "Badischer Volksbote", July 1, 1893

This question is the most important in our national affliction of drouth, and the lack of pasturage connected therewith, that can occupy any true friend of his country. And this question is not answered by old party catchwords of protective tariff and free trade and monopoly; it will neither be solved in the Reichstag nor in the local legislature, though legislation is also in this matter an important factor. The farmer alone can decide here; in his hand lies the future of our people. The matter at stake is the most valuable possession a people can have, their native land and soil. These are faring ill. Our land is not only being more heavily encumbered with mortgages every year, but is also losing some of its good qualities and fertility, and as the debt increases the value decreases. This is the most threatening complication we have to meet. But there is no use in merely lamenting it, it must be improved and amended. And it can be improved if we will only open our eyes and see and learn and act in accordance with that which we learn.

We can improve the soil and make it fertile by using stone-meal as a fertilizer, as is shown by the experience of many practical farmers. In the "Neues Mannheimer Volksblatt", M. A. Heilig publishes the following declaration:

"The Landwirthschaftliche Blaetter," by Mr. Councillor Nessler, in Karlsruhe, rejected a few months ago Hensel's method of mineral manuring. Whoever wants to convince himself how Hensel's method acts in practice is invited to inspect my two and one-half acres of barley near the Isolating Hospital. Despite of the unusual drouth the barley has attained an unusual height, and stands much fresher than the barley in other fields. After the harvest I shall have the yield determined before witnesses to see the difference also in this respect".

When practical experiments show such results the farmer ought to give up his old prejudices and try himself whether the new method of manuring is not better than the old. That the scientists and professors ignore the new source of fertilizing need not astonish us, on the contrary: "The professors are opposed to it, therefore it is good", may soon become a proverb, for hitherto the professors have always opposed everything good at its first appearance. We think Hensel's method of manuring will likely make agriculture again profitable, and we shall recommend it even if all should oppose us on this account. When at some future date, not too far removed, the German farmer and through him all the German people shall enjoy the blessings of this improvement of the soil we shall yet receive thanks that we helped to prepare the path for this new good during its hard times.

FROM THE "RHEINISCHER COURIER"
Wiesbaden, June 6, 1893

We have received the following communication: "In No. 152 of your valued journal, among the "Agricultural Communications", is a short, but favorable notice, from the "Manure-Division" of the "German Agricultural Society", concerning stone-meal. With respect to this, permit me to invite you and every one interested to examine the fields and vineyards of my friend Franz Brodtmann here, as also the rye fields of Mr. Heil, the tenant farmer at Hof Steinheim, on the estates of the Grand Duke of Luxemburg, which had been manured with this material according to my directions, and they will be convinced that contrary to these views stone-meal is a most important fertilizer of the very best quality, which when rightly used yields the best results.

Respectfully,

L. FORKB.
Eltville, June 4, 1893

FROM THE "REINISCHER COURIER"
June 29, 1893

Communication No. 175 of your morning edition of June 26 contains an attack on stone-meal as a manure, and an exaltation of the present method of manuring with potash, nitrogen and phosphoric acid. I was for many years an adherent of this latter method, but I have become convinced by experience and practical trials that these artificial manures serve indeed to *force the growth* and may be used with effect for several years, but that they do not restore to the soil what we withdraw from it in cultivation. Therefore the state of our soil unavoidably deteriorates from year to year, and at last refuses its service. Nobody can stand partridges every day, but he can his daily bread, and so it is with all plants, which not only need potash, nitrogen and phosphoric acid for their nourishment, but in addition soda, lime, magnesia, sulphuric acid, silicic acid, chlorine, iron, fluorine, carbonic acid, etc. All these elements are found in many rocks in greater or smaller quantities, and Hensel cannot be sufficiently thanked that he has pointed out to us farmers these inexhaustible supplies. When we return stone-meal to the soil we restore to it all that was in the soil from the beginning, and that our early ancestors did well with the original material is manifest as stable manure has only been used for about 200 years, and so-called artificial manure only about 50years. Of course we cannot force matters with stone-meal; but if it is brought on the fields in autumn and plowed under we may count on success as may clearly be seen here and as I have already stated in No. 155 of your much valued paper. With all esteem for science, we farmers cannot be contented with simply finding out how much potash, nitrogen and phosphoric acid the artificial fertilizers contain and how much every % thereof costs, we must rather strive to raise good crops on our fields with slight expense, without at the same time causing our soil to deteriorate by a one-sided system of fertilizing, and this is certainly done when we only apply potash, nitrogen and phosphoric acid.

L. FORKE
Eltville, June 27, 1893

FROM THE "NEUES MANHHEIMER VOLKSBLATT"
July 19, 1893

That the much-abused stone-meal cannot be without its excellent points the results in the fields best show. Mr. Kircher here has raised on various fields manured with this material barley and wheat, which must absolutely convince even the most skeptical of the usefulness of this manure. First, not only are the stalks considerably higher and stronger than those from fields manured with other material, but the ears are on the average 33% longer and the grains considerably more perfect. (Mr. Kircher has left in the editorial room of the "N. M. V." several wheat ears and barley ears from his fields to show the difference, also some from neighboring fields which have not been manured with Hensel's fertilizer. Whoever is interested in this matter, and every farmer should be so, may inspect the ears in our office).

"IRON SLAG" FROM THE "KOELNISCHE VOLKSZEITUNG"
April, 1893, No. 234, First Sheet

The supplement of the "Thueringer Landboten" brings a noteworthy article by the practical farmer, A. Armstadt, under the heading: "The Future of the Iron Slag". The author first notes that iron slag has risen to be the most generally used fertilizer containing phosphoric acid only in consequence of an immense amount of advertising, but now it seems to be about to lose much of its reputation. Even the "German Agricultural Society" will earnestly declare against it in its next publication.

"I myself", says A. Armstadt, "have never been enabled to feel any enthusiasm for iron slag in consequence of my experiments with it. And I have frequently on various occasions declared this, and it is a satisfaction to me that numerous reports are now appearing which confirm my observations. First of all, the fact that people come to doubt the theory of a gradual enrichment of the soil thereby will cause it to lose credit.

Men of science, as is well known, gave out the notion that the soil must gradually be enriched with phosphoric acid in order that rich crops may be raised. Iron slag was said to be the most suitable for this purpose, not only because the phosphoric acid in it is cheapest, but also because phosphoric acid in this form would in time become more soluble. But most farmers have waited probably in vain for the after effects. I myself have never found any after effects. According to the latest experiments, it is not only probable but pretty well established that every enrichment of the soil with phosphoric acid in mineral form is a waste, for it passes into a form difficult of solution, so that it cannot any longer be taken up by plants".

Prof. Dr. Liebscher (Goettingen) even found that with a manuring of 100 cwt. (5.000 kg) of iron slag to 60% of an acre no after effects developed, though he waited for it for seven years. But the copious applications of iron slag are founded on this theory of enrichment only.

"NEUES MANNNHEIMER VOLKSBLATT"
August 3, 1893

With a few potted plants or a small piece of garden anyone can make a trial of the value or worthlessness of Hensel's teachings, and no more paper need be wasted in their justification. An increasing number of farmers are experimenting successfully with the new fertilizer and it will gradually but surely supersede the old. The old manure supplied plants with too much *forcing* material and too much phosphoric acid, a substance which surely causes plant-lice, caterpillars, snails and the like. The stone-meal improves the nutrition of the plants without forcing them, so that while their growth is slower their leaves have a lesser amount of water, the fruits and stalks a greater amount of lime and are more wholesome and nourishing. As the fruits mature the phosphorus passes mostly into the seeds and the silica into the leaves and stalks. When agriculture hitherto built its theory of manuring on the ashy constituents of the seeds with their high contents of phosphorus it did not consider that the whole growing plant before the separating process during ripening requires quite different proportions of admixture than what may be derived from the seeds alone. A comparison of Hensel's views on this domain with the question of human nutrition rises very naturally. Exhausted men also are favored with allowing them to eat heartily of the convenient meat, with eggs and milk, all nutriments fully prepared for assimilation. The consequence is an excitation and irritation of the whole organism, bad digestion, increased watery contents of the body, perspiration, thirst, exhaustion from slight exertions, debility. A strong manuring with predominantly animal offal is for plants planted in a soil deficient in certain minerals what a predominantly animal diet is for men. If we look at men who live in the country almost altogether on food difficult of assimilation, of bread, vegetables and fruit, we observe a far more quiet bodily activity, little perspiration, little thirst, great and continuous muscular power. It is similar with plants when we offer them again their original nutriments, direct them to the appropriation of mineral constituents and give them organic manures or nitrogen only in small quantities and as a secondary matter. In both cases the constitution will be more normal, freer from parasites (diseases). If we notice in agricultural journals the enormous expenditures for advertising artificial manures it may be known what a gain those factories yield, and the mind grows sad at the wealth withdrawn from German farmers, who even without this are so hard pressed.

DR. E. SCHLEGEL,
Pract. Physician, Tuebingen

"WIESBADENER GENERAL ANZEIGER", July 8, 1893

For diminishing the distress as to fodder, we do not need as the troubled farmer is advised in another journal, to use artificial manure: superphosphate and Chili-nitre or superphosphate of nitrate of potash for the meadows; superphosphate of nitre with acid phosphate or with phosphate of lime for the cloverfields; fresh stable manure and liquid manure, Chili-nitre, superphosphate of potash or superphosphate of nitre for Indian corn for the horses, etc. The pen and compositors object to the 20-fold repotition of the wonderful compounded fertilizers. We recommend for the meadows, ashes of every kind, and for the root fields, street dust, and in general for the future, mineral manure, which is at the same time the best protection against drouth and all the diseases of plants, as it gives to plants the power of resistance and which in turn is transferred to men and animals through their food. That the diminution of the present and also of any future distress as to fodder may be effected by mineral manure is demonstrated by the following experience:

For 5 years I have been using stone-meal manure in my garden and fields. The results always have been satisfactory in every respect, for the soil becomes better every year by using this manure. Especially this year during the extraordinary drouth, the excellent effects of stone-meal fully manifested themselves. The flower as well as the different vegetables developed so magnificently that every one who passed my garden stopped and admired the great growth, especially of the Kohlrabi. In the cabbage which I planted at the beginning of April in my cow pasture, the rich crop is the more astonishing as it was not watered during the whole of its growth, this field has received *for 5 years only stone-meal manure (no solid or liquid stable manure)*; alongside of the cabbage field is the potato patch and it shows a most luxurious growth *despite of the abnormal drouth*. The above experience has brought me to the firm conviction that this fertilizer not only improves and augments the cultivated soil, but also keeps it moist and therefore prevents the rapid drying up of plants during a drouth.

BERNH. WETTENGEL,
Horticulturist and Truck-Farmer, Frankenthal, July 1, 1893

For two years I have used stone-meal manure with the greatest success, and especially this year, *despite the extraordinary drouth*. The result has been magnificent; the barley showed a much larger yield of grain than ever before; the potatoes were very fine and to our astonishment remained *untouched by the heavy frosts*, though others that had received stable manure suffered very severely. I was very much pleased with the effects on oats and clover. *Quite astonishing also is the dark green, full leaved appearance of the sugar beets, notwithstanding the great continuous drouth*. With the fruit trees where I especially applied the new fertilizer, I have fully learned how extraordinarily it acts. I would therefore urgently recommend every farmer to adopt the new method. With the greatest satisfaction I sign myself

PETER HEILMANN,
Agriculturist, Moersch, near Frankenthal, June 30, 1893

DETAILLED STONE-MEAL-EXPERIMENT:

In order to determine the results of the new method of fertilizing, the undersigned farmers and friends of agriculture assembled on June 25, 1893, early in the morning at 7 o'clock sharp, for a common inspection of the fields, and at this occasion we inspected the following fields within the domain of Frankenthal:

Name of fields:	Planted with:	By the farmer:
Muehlgewann,	Potatoes,	Carl Hermann, I.
New Gardens,	Barley,	Conrad Bender, widow
Grosse Garkueche,	Barley,	Peter Huber, widow
Grosse Garkueche,	Rye,	Adam Mack, I.
Bohrlache,	Barley,	Daniel Scherr
Kleiner Wald,	Barley,	Valt. Zimmermann
Kuhweide,	Potatoes, cabbage	Bernhard Wettengel
Schiesshaus,	Barley,	The rifle club
Actien-Eiskeller,	Potatoes,	The stock company
Gartengewann 1 Wormserstr,	Barley,	Clem. Wurmser
Gartengewann 2 Wormserstr,	Rye,	Wilh. Schwarz
Gartengewann 3 Wormserstr,	Barley,	Jah. Mees
Erbbestand,	Barley,	Hen. Grueming
Gartengewann 4 Wormserstr,	Barley,	Phil. Schatz
1 Mittelgewann,	Barley,	Joh. Bender, widow
Spiegelgewann,	Barley,	Valt. Zimmermann
Wingertsgewann,	Potatoes,	A. Gensheimer
Wingertsgewann,	Potatoes,	Jac. Armbrust, fieldguard
Neuweide,	Sugar-beets,	Pet. Diehl, Beindersheim

Neuweide,	Sugar-beets,	Conr. Peters, Beindersheim
Pfaffengewann,	Barley,	L. Braunsberg, II.
Pfaffengewann,	Potatoes,	Phil Schatz

Nearly all taking part in the inspection were practical farmers, who are entirely familiar with the local relations and quality of the fields. The result of the inspection may well be called astonishing.

Though the summer has been abnormally dry, all the barley inspected was distinguished by its dark green appearance when compared with other fields not fertilized with stone-meal. The ears compared with the others contained more rows. In a number of them we counted fourty grains extraordinarily fine and well developed. The same conditions existed with the rye. The potato fields showed a surprising luxuriant stand. We must especially mention the full leaved dark green appearance of the sugar-beets, which encourages us to look forward to a full development of the roots. With the cabbage the rich crop is the more surprising as it had not been watered during the whole period of its growth.

The undersigned have taken part in this general inspection with the more interest, as they are convinced that the violent dispute concerning the new method of manuring can only be decided by practical success. This was the reason why they desired to determine the numerous results obtained by a general local survey made in the above mentioned way in a conscientious manner, and they believe that they have thus ministered to the common good.

Biendersheim: P. Dichl; Edigheim: H. Jaeger, Jean Loosmann; Flowersheim: G. Garst, Ph. Schreiber; Frankenthal: J. Armbrust, Fr. Bender, J. Fries, J. Fueschsle, K. Gaschott, G. Kirchner, C. Luehle, H. Mayer, J. Mees, C. Moeller, C. Rupp, Ph. Scnatz, D. Scherr, Fr. Scheuermann, G. Wettengel, Jos. Zimmermann; Friesenheim: Chr. Moellinger; Moersch: P. Heilmann; Oppau: W. Claus.

END

𝔈mpfehlenswerte Schriften von Julius Hensel:

Julius Hensel.

Bezug im Internet (lulu.com, amazon, Barnes&Noble usw.), vorzugsweise direkt auf dem Blog **www.julius-hensel.com**:

„über causalmechanische Entstehung von Organismen"
von Pilgermann (Pseudonym Julius Hensel), Ersterscheinung 1881, aktuell überarbeitet und verlegt durch John Schacher bei lulu, Paperback, ISBN 978-1-4467-1396-9, Deutsch, 114 Seiten
Euro 12,90 *(e-book: Euro 4,90)*

„Makrobiotik oder unsere Krankheiten und
unsere Heilmittel" von Dr. Julius Hensel, Ersterscheinung 1882, aktuell gemäß der 2. Auflage überarbeitet und neu verlegt von John Schacher bei lulu,
Hardcover (Leinen/Schutzumschlag),
ISBN 978-1-4457-7900-3, Deutsch, 208 Seiten
Euro 24,90 *(e-book: Euro 9,90)*

„Makrobiotik oder unsere Krankheiten und unsere Heilmittel" von Dr. Julius Hensel, Ersterscheinung 1882, aktuell gemäß der 2. Auflage überarbeitet und neu verlegt von John Schacher bei lulu, Paperback, ISBN 979-1-4467-1408-9,
Deutsch, 208 Seiten
Euro 16,90 *(e-book: Euro 9,90)*

„Das Leben – seine Grundlagen und die Mittel zu seiner Erhaltung"
von Dr. Julius Hensel, Ersterscheinung 1885, aktuell überarbeitet und neu verlegt durch John Schacher bei lulu.com,
Hardcover (Leinen/Schutzumschlag),
ISBN 978-1-4461-3277-7, Deutsch, 520 Seiten
Euro 34,90 *(e-book: Euro 14,90)*

„Brot aus Steinen – ein neues und rationelles System zur Felddüngung und körperlichen Regeneration"
von Dr. Julius Hensel, Ersterscheinung 1892. Aus dem Englischen von John Schacher, neu verlegt bei lulu.com
Paperback, ISBN 978-1-4476-6803-9, Deutsch, 129 Seiten
Euro 12,90 *(e-book: Euro 4,90)*

„Physiologisches Brot"
von Dr. Julius Hensel, Ersterscheinung 1894. Aus dem Englischen von John Schacher, neu verlegt bei lulu.com
Paperback, ISBN 978-1-4476-2129-4, Deutsch, 38 Seiten
Euro 7,90 *(e-book: Euro 4,90)*

„Vereinfachte Heilkunst – Rheuma & Tuberkulose – wie entstehen Bazillen?" von Dr. Julius Hensel, Ersterscheinung 1899, aktuell überarbeitet und neu verlegt von John Schacher bei lulu.com,
Paperback, ISBN 978-1-4476-2937-5, Deutsch, 72 Seiten
Euro 12,90 *(e-book: Euro 7,90)*

*„Die lebenswichtige Bedeutung der Mineralstoffe
des Blutes und der gesamten Leibessubstanz"*
von Dr. Julius Hensel, 1904, aktuell überarbeitet und verlegt von
John Schacher bei lulu.com,
Paperback, ISBN 978-1-4476-2938-2, Deutsch, 38 Seiten
Euro 7,90 *(e-book: Euro 4,90)*

*„Das Wichtigste von der ganzen Heilkunst oder
was braucht der Mensch zum Leben und Gesundbleiben?"*
von Dr. Julius Hensel, Ersterscheinung 1904, aktuell überarbeitet
und neu verlegt von John Schacher bei lulu.com,
Hardcover/Schutzumschlag, ISBN 978-1-4467-2940-5,
Deutsch, 160 Seiten
Euro 24,90 *(e-book: Euro 9,90)*

englische Ausgaben:

„Bread from Stones"
by Dr. Julius Hensel 1892, new edition by John Schacher,
Paperback, ISBN 978-1-4467-5966-0, english, 96 pages
Euro 9,90 *(e-book: Euro 4,90)*

„Physiological Bread"
by Dr. Julius Hensel 1894, new edition by John Schacher,
Paperback, ISBN 978-1-4476-2131-7, english, 36 pages
Euro 7,90 *(e-book: Euro 4,90)*

Made in the USA
Middletown, DE
22 June 2024

56156899R00061